Sanitation in unsewered urban poor areas:

technology selection, quantitative microbial risk assessment

and grey water treatment

Alex Yasoni Katukiza

Thesis committee

Promotor
Prof. Dr P.N.L. Lens
Professor of Environmental Biotechnology
UNESCO-IHE, Delft

Co-promotor
Dr M. Ronteltap
Lecturer in Sanitary Engineering
UNESCO-IHE, Delft

Other members
Prof. Dr G. Zeeman, Wageningen University

Em. Prof. Dr W. Verstraete, Ghent University, Belgium

Prof. Dr P.D. Jenssen, Norwegian University of Life Sciences, Ås, Norway

Dr T.O. Okurut, National Environmental Management Authority, Kampala, Uganda

This research was conducted under the auspices of the Graduate School for Socio-Economic and Natural Sciences of the Environment (SENSE)

Sanitation in unsewered urban poor areas:

technology selection, quantitative microbial risk assessment

and grey water treatment

Thesis
submitted in fulfilment of the requirements of
the Academic Board of Wageningen University and
the Academic Board of the UNESCO-IHE Institute for Water Education
for the degree of doctor
to be defended in public
on Friday, 29 November 2013
at 4 p.m. in Delft, The Netherlands

by

Alex Yasoni KATUKIZA
born in Kabale, Uganda

CRC Press/Balkema is an imprint of the Taylor & Francis Group, an informa business

© 2013, Alex Yasoni Katukiza

Published by:

CRC Press/Balkema

PO Box 11320, 2301 EH Leiden, The Netherlands

e-mail: Pub.NL@taylorandfrancis.com

www.crcpress.com – www.taylorandfrancis.com

ISBN 978-1-138-01555-5 (Taylor & Francis Group)

ISBN 978-94-6173-769-4 (Wageningen University)

Dedication

This thesis is dedicated to my children as a motivation for them to strive and achieve what they want in life.

To my wife whose love and patience provided the strength I needed to progress.

To my parents who valued education and enabled me to reach where they could not reach.

Acknowledgments

This research was funded by the Netherlands Ministry of Development Cooperation (DGIS) through the UNESCO-IHE Partnership Research Fund. It was carried out at UNESCO-IHE, Delft (The Netherlands) and Makerere University, School of Engineering (Uganda) in the framework of the research project 'Addressing the Sanitation Crisis in Unsewered Slum Areas of African Mega-cities' (SCUSA).

I wish to express my sincere gratitude to my promotor Prof. Dr. ir. Piet Lens and supervisors Dr. ir. Mariska Ronteltap, Prof. Dr. Frank Kansiime and Dr. Charles Niwagaba for their strong scientific support, guidance and encouragement throughout the PhD research period. I will never forget the continuous critical comments from Prof. Dr. Piet Lens, that seemed to cause tough times but in the end enabled me to be critical, open minded and carryout multiple tasks with tight deadlines. I would like to thank Dr. Mariska again for her encouragement and scientific guidance during tough times and for the assistance in translating the summary into Dutch. I look forward to future research collaboration with my supervisors.

I would like to thank the SCUSA project manager Assoc. Prof. Dr. Jan Willem Foppen for his scientific support and assistance during the course of the research. Thanks to my colleagues Philip Nyenje and John. B. Isunju with whom I passed through the hard times and enjoyed good times during the course of this research. I thank Fred Kruis, Peter Heerings and Lyzette Robbemont of UNESCO-IHE for their assistance in obtaining research equipment and laboratory consumables and Jolanda Boots for her assistance in transfering funds and obtaining airtickets. Thanks go to Dr. Kulabako Robinah for assistance with laboratory equipment and encouragement. The support provided by John Omara and Rita Nakazibwe during analysis of samples and by Fred Mukasa during construction of the filtration systems for the grey water treatment is highly appreciated. The following carried out their MSc. research within the framework of this SCUSA project 1 and their contribution is appreciated: Albert Oleja, Hayeloum Temanu, Olivier Goldschmidt and Henrietta Osei-Tutu.

I thank my wife Joan and our beloved children Jethro Mugume and Ashley Abindabyamu for their love, support, prayers and also patience during my absence while abroad. To my parents and siblings, in-laws and close friends; thank you for your prayers, support and love.

Alex Y. KATUKIZA

Delft, 29 November 2013

Abstract

The sanitation crisis in unsewered urban slums of cities in developing countries is one of the challenges that need to be addressed. It is caused by the high rate of urbanisation in developing countries and the increasing urban population with limited urban infrastructure. The major issues of concern are the collection, treatment and safe disposal of excreta, grey water and solid waste. The goal of this study was to contribute to the sanitation improvement in urban slums with focus on sanitation technologies.

A review of sanitation technology options for urban slums was made followed by a baseline study in the slum of Bwaise III in Kampala Uganda. The results from the situation assessment and analysis were used to develop a method for selection of sustainable sanitation technologies in urban slums. Quantitative microbial risk assessment was then carried out based on the sources and concentration of pathogens and indicator organisms in the slum environment. The risk of infection and the disease burden contribution from various exposure pathways were determined. The study then focused on grey water treatment using a low-cost media (sand, crushed lava rock) based systems at laboratory scale and household level in the study area.

The results showed that existing facilities in Bwaise III are unimproved and do not function as elements within a sanitation system. In addition, there is no system in place for grey water management. There was also wide spread viral and bacterial contamination in the area. The maximum concentration of human adenoviruses F and G (HAdV-F and G) rotavirus (RV) was 2.65×10^1 genomic copies per mL (gc mL^{-1}) and 1.87×10^2 gc mL^{-1}, respectively. The concentration of *Escherichia coli* and *Salmonella* spp. ranged from 3.77×10^4 cfu. (100 mL)$^{-1}$ to 2.05×10^7 cfu. (100 mL)$^{-1}$. The disease burden from each of the exposure routes in Bwaise III slum was 10^2 to 10^5 higher than the World Health Organisation (WHO) tolerable risk of 1×10^{-6} disability-adjusted life years (DALYs) per person per year. Grey water generated in Bwaise III amounted to 85% of the domestic water consumption and was highly polluted with a COD and TN concentration range of 3000-8000 mg.L^{-1} and 30-50 mg.L^{-1}, respectively, and *Escherichia coli (E. coli)* concentration of up to 2.05×10^7 cfu. (100 mL)$^{-1}$. Grey water treatment with a crushed lava rock filter and using a two-step filtration process, resulted in the COD and TSS removal efficiencies of 88% and 90%, respectively, at a constant Hydraulic Loading rate (HLR) of 0.39 m.d^{-1} In addition, the highest removal efficiencies of TP and TKN were 59.5% and 69%, respectively, at a HLR of 0.39 m.d^{-1}. A log removal of *E. coli*, *Salmonella* spp. and total coliforms of more than 3 (99.9%) was also achieved under household filter usage conditions.

These results show that grey water treatment using a two-step crushed lava rock filter at household level in an urban slum has the potential to reduce the grey water pollutant loads by 50 % to 85%. However, its impact on public health and the environment needs to be assessed after its wide application. The need for advanced removal of pathogens and micro-pollutants from grey water warrants further research. In addition, the management systems for other waste streams of excreta and solid waste need to be in place as well to achieve the desired health impacts in urban slums. Integration of quantitative microbial risk assessment (QMRA) in the selection process of sustainable sanitation technologies for urban slums is recommended for future studies aimed at providing a holistic approach for upgrading slum sanitation. This will help to further understand the health impacts and benefits of sanitation solutions and also provide support to local authorities in making decisions on the measures to reduce the disease burden and environmental pollution.

Contents

Chapter 1: General introduction

1.1 Sanitation in urban slums of developing countries

The rates of urbanisation and urban slum growth in developing countries especially in sub-Saharan Africa, South America and Asia are estimated to be increasing and higher than the rate of urban infrastructure and services provision (Isunju et al., 2011; WHO and UNICEF, 2012). Urban slums are characterised by high population density, population dynamics, poor urban infrastructure and lack of legal status (Katukiza et al., 2010). These factors make the provision of sustainable sanitation services difficult, which has also led to the increase in the urban population without access to improved sanitation in major urban centres in developing countries (Cairncross, 2006; WHO and UNICEF, 2012). In addition, the funds budgeted for the water and sanitation sector for example are mainly spent on water supply infrastructure, which has further weakened the sanitation sub-sector leading to the sanitation targets not met by most developing countries (Moe and Rheingans, 2006; Joyce et al., 2010).

Generally, inadequate collection and treatment of the waste streams (excreta, grey water and solid waste) and safe disposal or reuse of the end products is a threat to the environment and a risk to public health. In urban slums, soil and water sources (such as boreholes, shallow wells, springs and streams) are contaminated with pathogens (bacteria, viruses), nutrients (NO_3^-, PO_4^{3-}, NH_4^+) and micro-pollutants (Howard et al., 2003; Katukiza et al., 2013; Nyenje et al., 2013). In particular pit latrines in slums contaminate ground water sources (Graham and Polizzotto, 2013; Nyenje et al., 2013), which may have negative health impacts on the slum dwellers. Moreover, high child mortality rate and loss of working days as a result of morbidity in urban poor areas are attributed to inadequate sanitation and poor hygiene practices (Genser et al., 2008; Rutstein, 2000). Provision of adequate and improved sanitation in slums is thus driven by the need to improve the quality of life by protecting the exposed population from infectious diseases, to reduce deterioration of water sources, to protect the ecosystem downstream the urban slums and to recover waste for economic benefits in the form of renewable energy, reclaimed water and recyclable solid materials.

The dominant type of sanitation facilities in urban slums in developing countries is mainly pit latrines used for excreta disposal (Thye et al., 2011; Howard et al., 2003). They require low capital and operating costs, are non-waterborne and can be easily built and maintained locally. Pit latrines are usually elevated in high water table areas (Katukiza et al., 2010). The high filling rate due to higher user-load and disposal of non-biodegradable solids in the pit latrine chamber is a challenge to the sustainability of pit latrines in urban slums. In addition, there is lack of access for pit emptying with cesspool emptiers whose cost may not be affordable by the slum dweller. Manual pit latrine emptying from the chamber to the adjacent excavated hole is therefore commonly practiced because it is the cheapest option, despite its negative health and environmental consequences. Alternative options in form of Vacutug MK1, Vacutug MK 2 and the MAPET have been used in some parts of Africa and Asia (Thye et al., 2011). Sanitation technology innovations in form of urine diversion dehydrating toilet

(UDDT), community sanitation blocks, Sulabh flush compost toilet and biogas toilets have also been implemented in Asia and Africa with the aim of improving sanitation in slums. They provide additional benefits in form of biogas and manure or soil conditioner. However, there are still questions on the categorisation of sanitation facilities as improved and unimproved by the Joint Monitoring Program (JMP) of UNICEF and the World Health Organization (WHO) based on technology approach rather than function based approach (Kvarnström et al., 2011). Moreover, this categorisation by UNICEF and WHO needs to include sanitation technologies for management of solid waste and grey water as well.

Simplified sewerage has been implemented for off-site treatment of combined sewage and grey water in South Africa, Sri Lanka, Brazil and other countries in the same regions (Mara, 2003; Paterson et al., 2007). Although it is considered cheaper based on the economies of scale (Paterson et al., 2007), its feasibility in densely populated urban slums is hampered by limited space, low affordability for waterborne systems and lack of reliable piped water supply. Off-site treatment of excreta and grey water does not offer opportunities for source separation of the waste and resource (in form of nutrients and energy) recovery. It is therefore critical to be able select appropriate technologies for a given geographical location or practical situation and to make technologies function within a system and acceptable by the beneficiaries. In addition, sustainability of sanitation systems is affected by inter-linked technical and non-technical factors including institutional arrangements for up-scaling and replication by practitioners (Jenkins and Sugden, 2006).

1.2   Research scope and objectives

This study was carried out in the framework of the interdisciplinary research project SCUSA (Sanitation Crisis in Unsewered Slum Areas in African mega-cities). It was comprised of three PhD sub-projects of Sanitation technologies (this research), hydrology and socio-economic aspects of sanitation in urban slums. The aim of the SCUSA project was to contribute to sanitation improvement in urban slums by integrating the technical, socio-economic and hydrological aspects of sanitation in slums. The study area of the SCUSA project was Bwaise III in Kampala (Uganda).

The specific objectives of this study based on the aim of the SCUSA research project were:
- To assess the sanitation situation in an urban slum of Bwaise III in Kampala (Uganda) and develop a method for selection of sustainable sanitation technologies.
- To provide an insight of the magnitude of microbial risks to public health caused by pathogens through various exposure pathways in typical urban slums such as Bwaise III in Kampala (Uganda).

- To design, implement and evaluate the performance of a grey water treatment technology (prototype) in an urban slum.

1.3   Thesis outline

The thesis consists of nine chapters. This first chapter gives a brief introduction of the study. Chapter 2 is based on literature review of technologies for urban slums and Chapter 3 presents a method for Selection of sustainable sanitation technologies for urban slums based on a baseline study in Bwaise III in Kampala (Uganda). Chapter 4 shows the results of genomic copy concentrations of selected waterborne viruses, while in Chapter 5 the magnitude of microbial risks from waterborne pathogens in a typical urban slum of Bwaise III in Kampala (Uganda) are presented. Chapters 6, 7 and 8, respectively, deal with the grey water characterisation and pollutant loads, laboratory-scale grey water treatment with a filter system and application of a two-step crushed lava rock filter system for grey water treatment at household level in the study area. The last chapter consists of general discussion, conclusions and recommendations for future research.

References

Cairncross S., 2006. Sanitation and water supply: practical lessons from the decade. UNDP – World Bank Water and Sanitation Program, The International Bank for Reconstruction and Development/The World Bank, Washington DC.

Carden, K., Armitage, N., Winter, K., Sichone, O., Rivett, U., Kahonde, J., 2007. The use and disposal of grey water in the non-sewered areas of South Africa: Part 1- Quantifying the grey water generated and assessing its quality. Water SA 33(4), 425-432.

Genser, B., Strina, ., dos Santos, L.A., Teles, C.A., Prado, M.S., Cairncross, S., Barreto, M.L., 2008. Impact of a city-wide sanitation intervention in a large urban centre on social, environmental and behavioural determinants of childhood diarrhoea: analysis of two cohort studies. International Journal of Epidemiology 37(4), 831 - 840.

Graham. J.P., Polizzotto, M.L., 2013. Pit latrines and their impacts on groundwater quality: a systematic review. Environmental Health Perspectives 121(5), 521-30.

Holm-Nielsen B, Al Seadi T, Oleskowicz-Popiel P. The future of anaerobic digestion and biogas utilization. Bioresource Technology 100 (22), 5478-5484

Howard, G., Pedley, S., Barret, M., Nalubega, M., Johal, K., 2003. Risk factors contributing to microbiological contamination of shallow groundwater in Kampala, Uganda. Water Research 37, 3421–9.

Isunju, J.B., Schwartz, K., Schouten, M.A., Johnson, W.P., van Dijk, M.P., 2011. Socio-economic aspects of improved sanitation in slums: A review. Public Health 125, 368-376.

Jenkins, M.W., Sugden, S., 2006. Rethinking sanitation. Lessons and innovation for sustainability and success in the New Millennium. UNDP - sanitation thematic paper.

Jingura, R.M, Matengaifa, R., 2009. Optimisation of biogas production by anaerobic digestion for sustainable energy development in Zimbabwe. Renewable and Sustainable Energy Reviews 13, 1116-1120.

Joyce, J., Granit, J., Frot, E., Hall, D., Haarmeyer, D., Lindström A., 2010. The Impact of the Global Financial Crisis on Financial Flows to the Water Sector in Sub-Saharan Africa. Stockholm, SIWI.

Katukiza, A.Y., Ronteltap, M., Niwagaba, C., Kansiime, F., Lens, P.N.L., 2010. Selection of sustainable sanitation technologies for urban slums - A case of Bwaise III in Kampala, Uganda, Science of the Total Environment 409(1), 52-62.

Katukiza, A.Y., Temanu H, Chung JW, Foppen JWA, Lens PNL., 2013. Genomic copy concentrations of selected waterborne viruses in a slum environment in Kampala, Uganda. Journal of Water and Health 11(2), 358-369.

Kulabako, N. R., Ssonko, N.K.M, Kinobe, J., 2011. Greywater Characteristics and Reuse in Tower Gardens in Peri-Urban Areas – Experiences of Kawaala, Kampala, Uganda. The Open Environmental Engineering Journal 4, 147-154.

Kvarnström, E., McConville, J., Bracken, P., Johansson, M., Fogde, M., 2011. The sanitation ladder – a need for a revamp?. Journal of Sanitation and Hygiene for Development 1(1), 3-12.

Mara, D.D., 2003. Water, sanitation and hygiene for the health of developing nations. Public Health 117(6), 452-456.

Moe, C.L, Rheingans, R.D., 2006. Global challenges in water, sanitation and health, Journal of Water and Health Suppl (04), 41-57.

Morel, A., Diener, S., 2006. Greywater Management in Low and Middle-Income Countries. Review of different treatment systems for households or neighbourhoods. http://www.eawag.ch/forschung/sandec/publikationen/ewm/dl/GW_managem ent.pdf [Accessed on 3rd January, 2013].

Nyenje, P.M., Foppen, J.W., Kulabako, R., Muwanga, A., Uhlenbrook, S., 2013. Nutrient pollution in shallow aquifers underlying pit latrines and domestic solid waste dumps in urban slums. Journal of Environmental Management 122, 15-24.

Okot-Okumu, J., Nyenje, R., 2011. Municipal solid waste management under decentralisation in Uganda. Habitat International 35, 537-543.

Paterson, C., Mara, D., Cutis, T., 2007. Pro-poor sanitation technologies. Geoforum 38(5), 901-907.

Rutstein, S.O., 2000. Factors associated with trends in infant and child mortality in developing countries during the 1990s. Bulletin of the World Health Organization. 78(10), ISSN 0042-9686.

Sall, O., Takahashi, Y., 2006. Physical, chemical and biological characteristics of stored greywater from unsewered suburban Dakar in Senegal. Urban Water Journal 3(3), 153-164.

Thye, Y.P., Templeton, M.R., Ali, M., 2011. A Critical Review of Technologies for Pit Latrine Emptying in Developing Countries, Critical Reviews in Environmental Science and Technology 41(20), 1793-1819.

UN, 2007. United Nations Millennium Development Goals Report. UN Statistics Division.

WHO and UNICEF., 2012. Progress on drinking water and sanitation. Joint Monitoring Program Report (JMP). 1211 Geneva 27, Switzerland.

Chapter 2:   Sanitation technology options for urban slums

This chapter is based on:

Katukiza, A.Y., Ronteltap, M., Niwagaba, C.B., Foppen, J.W.A., Kansiime, F., Lens, P.N.L., 2012. Sustainable sanitation technology options for urban slums. *Biotechnology Advances* 30, 964-978.

Abstract

Poor sanitation in urban slums results in increased prevalence of diseases and pollution of the environment. Excreta, grey water and solid wastes are the major contributors to the pollution load into the slum environment and pose risk on public health. The high rates of urbanization and population growth, poor accessibility and lack of legal status in urban slums make it difficult to improve their level of sanitation. New approaches may help to achieve the sanitation target of the Millennium Development Goal (MDG) 7; ensuring environmental sustainability. This paper reviews the characteristics of waste streams and the potential treatment processes and technologies that can be adopted and applied in urban slums in a sustainable way. Resource recovery oriented technologies minimise health risks and negative environmental impacts. In particular, there has been increasing recognition of the potential of anaerobic co-digestion for treatment of excreta and organic solid waste for energy recovery as an alternative to composting. Soil and sand filters coupled with a tertiary treatment step are suitable for removal of organic matter, pathogens, nutrients and micro-pollutants from grey water.

2.1   Introduction

2.1.1   Sanitation in slums

The need for sanitation improvement in urban slums is a result of unsanitary conditions and their negative effects on public health and the environment. Poor sanitation is part of the vicious circle of poverty and results in disease, illness and low productivity (Genser et al., 2008; Rutstein, 2000; Victora et al., 1988). In slums, human excreta (urine and faeces) are not properly managed. They are predominantly disposed by use of unlined pit latrines which are usually elevated to overcome periodic floods, ventilated improved pit latrines, flying toilets (use of polythene bags for excreta disposal that are dumped into the surrounding environment) or open defecation. In addition, solid waste is characteristically disposed of on illegal refuse dumps and grey water is discharged into open storm water drains or in the open space often resulting in ponding (Katukiza et al., 2010; Kulabako et al., 2010; Schouten and Mathenge, 2010).

Providing sanitation solutions accepted by the population living in urban slums is very challenging. It is hampered by: i) poor accessibility, which makes it difficult for cesspool emptiers and solid waste collection trucks to reach the area; ii) the lack of legal status of the area; slums typically arise from encroachment on land owned by the government and house owners are not willing to invest in permanent structures that may be demolished at any given time; and iii) the lack of interest in investing in sanitation facilities by inhabitants who are typically renting rather than owning the houses (Katukiza et al., 2010; Maksimović and Tejada-Guibert, 2001; Paterson et al., 2006, 2007). The growth dynamics of the urban slums over the last 15 years has indeed been unprecedented. Minor investments in improved sanitation have not been able to reduce the percentage of the urban unserved and this percentage is still expected to further rise (Figure 2.1). This is attributed to rural-urban migration and the low priority given to sanitation by urban authorities. The budget allocated to the sanitation subsector has been relatively low (Isunju et al., 2011; Okot-Okumu and Nyenje, 2011) although there has been considerable annual investment in the water and sanitation sector (Joyce et al., 2010). Sanitation investments are seen as large investments that are avoided, and the attention is more focused on water supply projects (Cairncross, 1992; Fang, 1999; Moe and Rheingans, 2006). As much as sanitation is seen as the corner stone of public health (WHO and UNICEF, 2010), the progress with respect to the meeting MDG sanitation target is hampered by low coverage of improved toilets and pit latrines in urban slums.

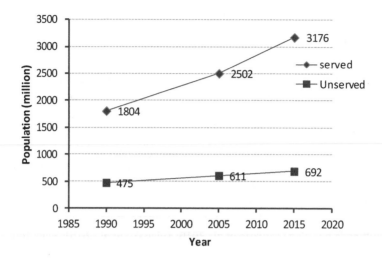

Figure 2.1: Global trends in improved sanitation of urban areas (WHO and UNICEF, 2006)

2.1.2 Public health and consequences of poor sanitation

Public health concerns associated with urbanization and slums include prevalence of vector borne diseases and bacterial infections, drinking water contamination and reuse of contaminated waste products (Mara et al., 2010; Nakagawa et al., 2006; Sidhu and Toze, 2009; Styen et al., 2004). Diseases such as cholera, dysentery, diarrhoea and malaria occur in slum areas as a result of poor sanitation and the presence of breeding areas for flies and mosquitoes (Crites and Tchobanoglous, 1998; Jha, 2003; Leeming et al., 1998; Maksimović and Tejada-Guibert, 2001). These poor conditions are reflected in the drinking water quality.

Boreholes, shallow wells and springs where the urban poor draw water from were found to be contaminated with Escherichia coli and viruses, and act as exposure routes for pathogens via drinking water (Ashbolt, 2004; Howard et al., 2003; Langergraber and Muellegger, 2005). The deterioration of the quality of groundwater and surface water resources is attributed to contamination from organic matter, micro-pollutants, nutrients and pathogens (Clara et al., 2005; Nakagawa et al., 2006; Nyenje et al., 2010; Foppen, 2002; Styen et al., 2004). Contamination of water resources, soil and food combined with poor hygiene practices such as lack of hand washing with soap have been the biggest cause for the high child mortality and loss of working days (Genser et al., 2008; Rutstein, 2000; Victora et al., 1988). Provision of sanitary facilities and proper hygiene practices complement each other in improving the well-being of the slum dwellers.

2.1.3   Sustainable sanitation

Providing sustainable sanitation for urban slums needs to address not only technology implementation but also cost, ownership and space issues. Sustainability with respect to sanitation implies the system needs to comprise of collection, storage, transport, and treatment of human excreta, grey water, solid waste and storm water, and the safe disposal or reuse of end products (Mara et al., 2011; Bracken et al., 2005; Kvarnström et al., 2004; Mara et al., 2007). A sustainable sanitation system should be technically feasible, acceptable to the users, affordable and contribute to health improvement and environmental protection. Population density, settlement pattern, landscape, water availability, household incomes, ownership and socio-cultural issues are also key factors that cannot be ignored (Avvannavar and Mani, 2008; Mara, 2008). Sustainability of sanitation also requires institutional structures and arrangements to be in place for operation, maintenance and up-scaling of interventions (Franceys, 2008; Jenkins and Sugden, 2006).

In urban slums, sanitation can be considered sustainable if it is able to sanitize waste for pathogen destruction, to protect ground water from pollution, to be implemented, operated and maintained at low cost and to function without use of materials of higher quality than necessary (Howard et al., 2003; Kvarnström et al., 2004; Ngoc and Schnitzer, 2009; Ottoson et al., 2003). Stakeholders' participation in the choice of sanitation technologies is key for sustainability although the level of affordability in slums is low (Katukiza et al., 2010), and a multi-disciplinary approach is essential (Mara et al., 2011). One way of keeping the total cost down is to aim for sanitation systems that provide additional income, such as renewable energy, reclaimed water and recyclable solid materials. Recyclables are most easily recovered when the different waste streams are kept separated; separation concentrates risks in small volumes and allows better control and limits negative environmental impacts (Kujawa-Roeleveld and Zeeman, 2006). Technologies that promote resource recovery using on-site sanitation systems are applied within the concepts of ecological sanitation (Ecosan) and decentralized sanitation and reuse (DeSaR). Typically very high cost sewer systems and dependency on the presence of running water can be avoided, thus DeSaR and Ecosan based technologies are considered to be accessible to the poor and fit in a system contributing to the MDGs (Bracken et al., 2007; Langergraber and Muellegger, 2005; Lettinga et al., 2001; Otterpohl et al., 1999; Paterson et al., 2006, 2007).

Sanitation contributing to attaining the MDGs suffers from one major difficulty; any sanitation facility shared by more than one household is considered to be unimproved by WHO (2010). An improved sanitation facility is defined by WHO as one that hygienically separates human excreta from human contact and not shared by more than one household (Table 2.1). Limited space makes non-shared household sanitation virtually impossible in a typical urban slum. It is common to find a building block with separate rooms rented out to at least10 households by a landlord who only provides a single stance pit latrine without any consideration of the high user-load (Katukiza et al., 2010). The suitability of all existing technologies in providing sustainable sanitation in urban slums has neither been examined nor has it been established which and how recent process innovations for resource recovery from urban waste streams can be made feasible for application in slums. This chapter reviews the state of knowledge with regard to waste characteristics, processes for pollutant removal from waste streams and the role of technology and technological innovation in solving the sanitation crisis in slums in a sustainable way.

Table 2.1:   Categorisation of sanitation facilities (WHO and UNICEF, 2010)

Improved sanitation facilities	Unimproved sanitation facilities
Flush or pour flush system to piped sewer system, septic tank, pit latrine	Flush/Pour flush to elsewhere
Ventilated improved pit (VIP) latrine	Pit latrine without slab/open pit
Pit latrine with slab	bucket latrine
Composting toilet	Elevated toilet/latrine
Urine Diversion Dehydrating (UDD) toilets	Public facilities
	Shared facilities by two or more households
	No facilities, bush or field, "flying toilets"

2.2 Waste streams in urban slums

The major waste streams into the environment in urban slums are excreta, grey water and solid wastes (Howard et al., 2003; Kulabako et al., 2007; Paterson et al., 2006, 2007). Waste streams have varying characteristics depending on the source and interaction with the environment. Pollution in slums is a result of pathogens, nutrients, micropollutants and other trace organics in waste streams (Genser et al., 2008; Ottoson and Stenström, 2003; Paterson et al., 2007; Styen et al., 2004). Understanding the components of the waste streams is important in applying technology to overcome the unique challenges in the urban slum environment.

2.2.1 Excreta

Excreta refer to urine and faeces. They form a major health risk due to the presence of pathogens in faeces and the mobility of nutrients and micro-pollutants in urine (Feachem et al., 1983; Prüss et al., 2002). In urban slums, where open defecation and excreta disposal in the open storm water drains is very common, this health risk is a reality (Chaggu et al., 2002; Paterson et al., 2007). Even when sanitation services are provided, the health risk is not eliminated. Pit latrines are designed to let the liquids percolate into the soil. In the densely populated areas, pollution of soil and groundwater is therefore another major public health risk. The major contaminant constituents in sludge from pit latrines and septic tanks are organic matter in the form of COD, nutrients and pathogens (bacteria, viruses and parasites). Their concentration is lower in the municipal wastewater treatment sludge compared to pit latrine sludge and septage as shown in Table 2.2.

Source separated human excreta have to be treated differently because of the different characteristics. Faeces has to be treated for removal of COD and destruction of pathogens, while urine has to be treated for removal of N and P. Urine has a larger fraction of nutrients in the form of nitrogen and phosphorus (Table 2.3). Pathogens are reported to be of faecal origin (Ottoson et al., 2003) and the level of sanitation in peri-urban areas greatly influences the bacteriological quality of drinking water from boreholes and springs (Ellis and Hvitved-Jacobsen, 1996). Pathogens of concern and indicator organisms for faecal contamination include enteric viruses, helminths (intestinal worms), bacteria, protozoa and faecal streptococci together with Escherichia coli. They are present in wastewater and biosolids contaminated with faecal matter (Sidhu and Toze, 2009).

Micro-pollutants primarily originate from the urine fraction of municipal wastewater. They form a particular threat to users of groundwater for drinking water in urban slums where the water table is mostly high and sanitation is poor. Accumulation of micro-pollutants in the environment leads to toxicity through the food chain and distortion of the ecological balance thereby causing environmental pollution (Clara et al., 2005; Joss et al., 2006). Pathogens present in excreta are presumed to be the major cause of disease outbreaks in urban slums although the long-term health effects of nutrient fluxes and release of micro-pollutants into the slum environment remain uninvestigated.

Table 2.2: Characteristics faecal sludge from onsite sanitation facilities and WWTP sludge

Parameter	Faecal and WWTP sludge characteristics (mean and range of values)			Source
	Public toilet sludge	Septage	WWTP sludge	
Total Solid, TS (mg/L)	52,500	12,000 -35,000	-	Koné and Strauss (2004)
	30,000	22,000		NWSC (2008)
	≥ 3.5%	< 3%	< 1%	Heinss et al. (1998)
Total Volatile solids (% TS)	68	50 - 73	-	Koné and Strauss (2004)
	65	45		NWSC (2008)
COD (mg/L)	49,000	1,200 -7,800	-	Koné and Strauss (2004)
	30,000	10,000	47-608	NWSC (2008)
	20,000 – 50,000	< 10,000	500–2,500	Heinss et al. (1998)
BOD$_5$ (mg/L)	7,600	840 - 2,600	-	Koné and Strauss (2004)
			20-229	NWSC (2008)
Total Nitrogen, TN (mg/L)	-	190 -300	-	Koné and Strauss (2004)
			32-250	NWSC (2008)
Total Kjeldahl Nitrogen, TKN (mg N/L)	3400	1,000		
NH$_4$-N (mg/L)	3,300	150 -1200	-	Koné and Strauss (2004)
	2,000	400	2-168	NWSC (2008)
	2,000 – 5,000	< 1,000	30–70	Heinss et al. (1998)
Nitrates, NO$_3^-$ (mg N/L)	-	-	-	NWSC (2008)
	-	-	-	Koné and Strauss (2004)
Total Phosphorus, TP (mg P/L)	450	150	9–63	NWSC (2008)
Faecal coliforms (cfu/100mL)	1x10^5	1x10^5	6.3x10^4 - 6.6x10^5	NWSC (2008)
Helminth eggs	25,000	4,000 - 5,700		Heinss et al. (1994)
	20,000–60,000	4,000	300–2,000	Heinss et al. (1998)
		600 - 6,000		Ingallinella et al. (2002)

Table 2.3: Characteristics of faeces, urine and grey water: loading rates and concentration

Parameter	Loading rates				Concentration			
	Unit	Faeces	Urine	Greywater	Unit	Faeces	Urine	Greywater
Organic matter								
COD	g/(ca.d)	37-63	10.0 - 12.0	7-102	g/L	10–50	4–11	0.35-0.78
BOD$_5$	g/(ca.d)	14-33.5	5.0-6.0	26-28	g/L	n/a	4	0.21-0.45
Nutrients								
N	g/(ca.d)	0.3-2.0	3.6-16	0.1-1.7	g/L	1.8-14	1.8-17.5	6.7-40
P	g/(ca.d)	0.3-0.7	0.4-2.5	0.1-2.2	g/L	2.2-4.1	0.2-3.7	0.4-31
K	g/(ca.d)	0.24-1.3	2.0-4.9	0.2-4.1	g/L	2.15-4.1	0.7-3.3	8.8
S	g/(ca.d)	0.2	0.6-1.3	0.5-7.7	g/L	-	1.2-2.6	72
Microorganisms								
E. coli	cfu/(ca.d)	1×10^{8d2}	-	6.3×10^{9d1}	cfu/100mL			2×10^{0}-3.5×10^{5a}
								3.8×10^{6}-8.5×10^{7b1}
								1×10^{8}-1.4×10^{9b2}
								7.5×10^{3}-2.6×10^{5c}
Total coliforms	cfu/(ca.d)	-	-	-	cfu/100mL	-	-	2.3×10^{3}–3.3×10^{5}

Meinzinger and Oldenburg, 2009; Kujawa-Roeleveld and Zeeman, 2006; Eriksson et al., 2002; Larsen and Maurer, 2011; Vinnerås et al., 2005, [a]Jefferson et al., 1999, b1 and b2 for wet and dry seasons (Dallas and Ho, 2005), Li et al., 2003; d1 and d2 based on flow of 64.9 L/(ca.d) and faecal load of 65g/(ca.d) (Ottoson et al., 2003).

2.2.3 Grey water

Grey water is wastewater of domestic origin from bathroom, kitchen, and laundry use, excluding wastewater from the toilets. Grey water accounts for 65 % to 75% of the domestic water consumption in peri-urban areas of developing countries where the water consumption per capita ranges from 20 L/ca.d to 30 L/ca.d (Elmitwalli et al., 2003; Eriksson et al., 2002; Morel and Diener, 2006). The amount of water consumed can thus be used to estimate the grey water production in the urban poor areas or slums. It can be argued that the potentially negative impacts from grey water disposal are in urban slums with water supply services and where little or no consideration has been given to the planning and management of grey water.

Grey water contains contaminants of concern that include suspended solids, pathogens, nutrients, grease and also organic micro-pollutants (Table 2.4) from household chemicals and pharmaceuticals that may be present due to urine contamination (Eriksson et al., 2009; Elmitwalli and Otterpohl, 2007; Li et al., 2009). Kitchen grey water contains a higher level of COD and total suspended solids (TSS) than grey water from the bathroom and laundry (Li et al., 2009). It has a nutrient content close to the COD:N:P ratio of 100:20:1 (Metcalf and Eddy, 2003) while other streams of grey water have low concentrations of nitrogen and phosphorus. The nutrient content is a limiting factor in application of conventional biological treatment processes for grey water treatment (Jefferson et al., 2001). Typically, all grey water types have a good biodegradability as indicated by their BOD_5 to COD ratio which is close to 0.5 (Knerr et al., 2008; Li et al., 2009).

Grey water contains pathogens as a result of faecal contamination and the highly biodegradable organic matter content that promotes growth of pathogens (Christova-Boal et al., 1996; Hargelius et al., 1995). It may contain high numbers of Escherichia coli up to 1.5x108 CFU/100 mL from showering and washing of diapers (Christova-Boal et al., 1995; Hargelius et al., 1995; Li et al., 2009) even though it may be considered less polluted that sewage. The bacteriological quality of grey water fractions at source points in urban slums is comparable to that of mixed grey water conveyed by tertiary drains in informal settlements, which makes it unfit for reuse without treatment (Carden et al., 2007). All the separate fractions of grey water from informal settlements may contain high levels of *E. coli* (Carden et al., 2007; Kulabako, 2009; Ridderstolpe, 2007) although studies in industrialised countries have shown that it is only true for the bathing fraction (Eriksson et al., 2002; Hargelius et al., 1995; Surendran and Wheatley, 1998).

Another pollutant is phosphorus present in high concentrations due to detergents used in kitchens, especially in developing countries where phosphates in detergents have not yet been replaced with other ingredients (Meinzinger and Oldenburg, 2008). Grey water quantity and quality vary depending on living standards, population characteristics (customs, habits) and the sanitation level of service. Grey water is only considered suitable for non-potable use after treatment depending on the effluent quality and the re-use application (Eriksson et al., 2002, 2009; Lens et al., 1994; Jefferson et al., 1999). In urban slums, the uncontrolled release of grey water requires source control and treatment mainly for removal of COD, nutrients (N, P) and pathogens.

Table 2.4:   **Characteristics of grey water types**

Parameter (units in mg/L unless otherwise stated)	Christova-Boal et al. (1996)	Li et al. (2009)	Surendran and Wheatley (1998)	Li et al. (2009)	Christova-Boal et al. (1996)	Surendran and Wheatley (1998)	Li et al. (2009)	Siegrist et al. (1976)	Hargelius et al. (1995)	Li et al. (2009)	Gerba et al. (1995)
Types of grey water	Bathroom	Bathroom	Bathroom	Laundry	Laundry	Laundry	Kitchen	Kitchen	Bathroom and Kitchen	Mixed	Mixed
Turbidity (NTU)	60-240	44-375	92	50-444	50-220	108	298	-	-	29-375	15.3-78.6
TSS	-	7-505	631	68-465	-	658	134-1300	2410	-	25-183	-
pH	6.4-8.1	6.4-8.1	7.6	7.1-10	9.3-10	8.1	5.9-7.4	-	-	6.3-8.1	6.7-7.6
Electrical cond. (μScm^{-1})	82-250	-	-	-	190-1400	-	-	-	-	-	-
BOD$_5$	76-200	50-300	425	48-472	48-290	-	536-1460	1460	-	47-456	-
COD	-	100-633	-	231-2950	-	725	26-2050	-	-	100-700	-
TOC	-	-	104	-	-	110	-	880	-	-	-
Oil and grease	37-78	-	-	-	9.0-88	-	-	-	-	-	-
Total N	-	3.6-19.4	-	1.1-40.3	-	-	11.4-74	74	-	1.7-34.3	-
TKN	4.6-20	-	-	-	1-40	10.7	-	-	-	-	-
NH$_4$-N	<0.1-15	-	1.56	-	<0.1-1.9	1.6	-	6	-	-	-
NO$_3$-N	-	-	0.9	-	-	-	-	0.3	-	-	1.8-3.0
Total P	0.11-1.8	0.11->48.8	-	ND->171	0.062-42	-	2.9->74	74	-	0.11-22.8	-
PO$_4$-P	-	-	1.63	-	-	10.1	-	31	-	-	-
Total coliforms (cfu/100ml)	$500-2.4\times10^{7}$	$10-2.4\times10^{7}$	6×10^{6}	$200.5-7\times10^{5}$	$2.3\times10^{3}-3.3\times10^{5}$	7×10^{5}	$>2.4\times10^{8}$	-	-	$56-8.03\times10^{7}$	$10^{7.2}-10^{8.8}$
Faecal coliforms (cfu/100ml)	$170-3.3\times10^{3}$	$0-4.4\times10^{5}$	600	$50-1.4\times10^{3}$	$110-1.09\times10^{3}$	728	-	-	40×10^{6}	$0.1-1.5\times10^{8}$	$10^{5.4}-10^{7.2}$
E. coli (cfu/100ml)	-	-	-	-	-	-	-	-	236×10^{6}	-	-

2.2.4 Solid Waste

Solid waste management is one of the major problems to the environment and to public health faced by developing countries. The growth of cities in developing countries has resulted in growth of peri-urban areas that generate large amounts of organic solid waste (Figure 2.2B). However, this growth has not been matched with the provision of suitable solid waste management practices at affordable costs in terms of collection, transport and treatment or safe disposal (Chanakya et al., 2009; Okot-Okumu and Nyenje, 2011).

| (A) | (B) | (C) |

Figure 2.2: **(A) A UDD toilet in a slum area of Kampala, Uganda, (B) A solid waste dump and an elevated pit latrine in the vicinity of a household in a slum area in Kampala (Uganda), (C) A flooded low-lying slum area of Bwaise III in Kampala (Uganda). This is after a storm event and it can stay like this for 2 days**

In developing countries, less than 50% of the solid waste generated in urban areas is collected centrally by the municipalities and private sector with limited recycling or recovery of recyclable materials (Jingura and Matengaifa, 2009; Kaseva and Mbulingwe, 2003; KCC, 2006; Okot-Okumu and Nyenje, 2011). This is a missed opportunity for resource recovery, economic benefits, and to reduce the waste quantities disposed in landfills (Kaseva & Gupta, 1996). The substantial amount of solid waste that remains uncollected is likely to result in environmental pollution and negative public health effects (Bhatia and Gurnani, 1996).

Solid waste generated in urban slums has a large portion that is organic (Table 2.5). Poor management of organic solid waste can lead to emission of methane from solid waste dumps as well as leaching of nutrients, micro-pollutants and organic matter to the natural environment (Holm-Nielsen et al., 2009; Sharholy et al., 2007). Uncollected solid waste also leads to blockage of storm water drains, thus increasing the flood risk in urban slums (Figure 2.2C). The increasing volume of solid waste generated with varying characteristics calls for a system approach from source separation to treatment options that promote resource recovery and finally disposal to a landfill of solely the inert fraction for which at the time no useful application can be found.

Table 2.5:   Composition of municipal solid waste in developing countries

Solid waste type	Composition (%)				
	Henry et al. (2006)	KCC (2006)	Ogwueleka (2009)	Rajabapaiah (1988)	Sharholy et al. (2007)
Food residues	53	74	76	65	45.3
Paper	16.8	11	6.6	8	3.6
Textile	2.6	1	1.4		2.2
Plastic	12.6	12	4	6	2.9
Grass/wood	5.6	-	-	7	
Leather	1	-	-	-	
Rubber	1.5	-	-	-	
Glass	2	2	3	6	0.7
Metal	2.3	0.4	2.5	3	2.5
Other waste	2.6	0.4	6.5	5	42.8[*]

*Includes leather, rubber, cardboard, wood, bricks, ash, soil

2.3 Sanitation systems for urban slums

2.3.1 Introduction

Treatment of waste streams in urban slums using sustainable technologies preserves their reuse potential to recover resources. Examples of resources are bio-energy generated from transformation of organic material, plant nutrients (nitrogen, phosphorus, potassium as macro nutrients; calcium and sulfur as micro-nutrients) and water treated to the desired discharge standards. Innovative waste disposal and treatment technologies have been developed in an attempt to address the sanitation crisis in urban slums, but they can only be sustainable if they function as elements within a sanitation system (Figure 2.3). A sanitation system is complete when it has a defined flow stream for each of the products (Tilley et al., 2010). Technologies and institutional aspects complement each other in the collection, transport, treatment and disposal or reuse of the waste. The settlement dynamics in urban slums have triggered new thinking because improved onsite sanitation will still play a key role in the foreseeable future in order to meet the MDGs (Lüthi et al., 2009; Mara and Alabaster, 2008; WHO and UNICEF, 2010). In urban slums there is no other way but to provide onsite shared sanitary facilities appropriate for the available limited space.

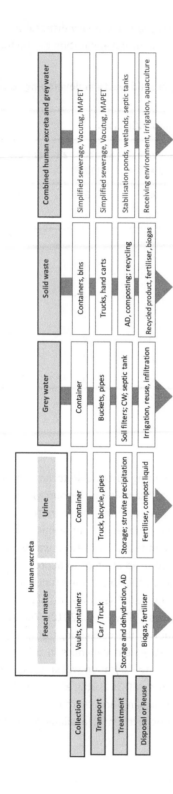

Figure 2.3: Interaction between elements of a sanitation system and domestic waste streams (AD: anaerobic digestion, CW: constructed wetlands) (GTZ, 2002; Li et al., 2009; von Münch and Mayumbelo, 2007; Tilley et al., 2008)

2.3.2 Collection and treatment of sewage

2.3.2.1 Off-site arrangement

Sewage here refers to combined excreta and grey water. It is collected and transported under gravity by a pipe network to the stabilization ponds for treatment before discharge into the receiving environment. Simplified sewerage is an alternative sanitation technology for densely populated urban settlements consisting of a sewer network and a wastewater treatment unit. High population densities result in high wastewater flows that make simplified sewerage attractive where there is lack of space for onsite sanitation (Paterson et al., 2007).

Simplified sewerage differs from conventional sewerage because it is characterized by reduced pipe diameters, gradients and depths without compromising the design principles (Bakalian et al., 1994). Cost savings in excavations, flexibility to lay pipes between housing blocks and under the pavements makes simplified sewerage suitable for urban slums (Paterson et al., 2007; Mara, 2003). It also allows for upgrading of the housing units with minimum relocation of existing structures and is considered to be cheaper than onsite sanitation based on economies of scale (Paterson et al., 2007). These properties have enabled implementation of simplified sewerage in per-urban areas in Pakistan, Sri Lanka, Brazil, Colombia, Peru, Bolivia and South Africa (Bakalian et al., 1994; Sinnatamby, 1990; Watson, 1995). Beneficiary communities have to be involved in the implementation, maintenance and operation of the simplified sewerage for sustainability (Mara, 2003).

Simplified sewerage is widely used in peri-urban areas worldwide but its application in urban slums is likely to meet the challenge of dynamic and illegal settlements resulting in low connections and a poor network. The importance of a multidisciplinary approach to engage communities and municipal authorities in project planning and implementation should not be overlooked. It raises awareness and subsequent adoption of a given sanitation technology. Investment in simplified sewerage has potential to serve urban slums to improve the health and quality of life, though it does not offer opportunities for waste separation and nutrient recovery.

2.3.2.2 On-site arrangement

A flush toilet connected to a septic tank that also receives grey water is an onsite system for sewage collection and treatment. The system requires water to transport the waste in pipes to a septic tank where it undergoes anaerobic treatment. The digested sludge is removed using a vacuum tanker after 3-4 years depending on the loading rate and the effluent is discharged in a soakaway before infiltration in the ground. Septic tank effluent can also be discharged into a constructed wetland. This technology is widely used in urban areas of developing countries where there is no sewerage (Mara and Alabaster, 2008; Tilley et al., 2008). In slums, it is used by the minority with house water connections because of the lack of ability to pay for the water, high cost of construction materials and space for the septic tank and soak way (Katukiza et al., 2010; Mara, 2007; Tilley et al., 2008). Community sanitation blocks using flush toilets and a septic tank have been implemented in urban poor areas of India and Africa (Burra et al., 2003; Mara and Alabaster, 2008). Their sustainability requires the strengthening of the legal and institutional framework for planning, implementation and operation and maintenance. This technology is appropriate for areas with low water table and not prone to flooding, and areas where there is space for a soak pit or constructed wetland and for managing grey water at the same time.

2.3.3 Collection and treatment of faecal sludge

2.3.3.1 Pit latrines

Pit latrines are the dominant type of excreta disposal facilities in urban slums in Africa, Asia and Latin America & Carribean (Chaggu, 2004; Howard et al., 2003; Kulabako et al., 2010; Mugo, 2006; Thye et al., 2011; UNICEF and WHO, 2010). The types of pit latrines in slums include traditional ones are made of wood poles and mud and ventilated improved pit latrines (VIP). Unlined pit latrines are usually elevated in slum areas with a high water table (Figure 2.2B). In slums, pit latrines are shared by many households as a result of high population density and limited space.

Pit latrines do not require water for their functionality, can be built and repaired with locally available materials, have low capital and operating costs and can be modified to serve user preferences (squatting, anal washing and wiping). Infiltration of the liquid phase into ground water and overflows during the rainy season from the excreta collection chamber have made pit latrines major causes of ground water pollution (Howard et al., 2003; Kulabako et al., 2007). High water table, flooding in rainy seasons and poor drainage most likely increase the risk of contamination of drinking water sources in urban slums.

The high user-load and filling rate of the pit latrines are a sanitation challenge in slums. Landlords are not held liable and the municipal councils are not under obligation to provide basic sanitation services in informal settlements that lack legal status (Dzwairo et al., 2006; Katukiza et al., 2010; Kulabako et al., 2010). In slums, where there is no accessibility for vehicles or where the owner cannot afford to pay for the service of a vacuum tanker, local contractors manually empty pit latrines by making a hole on the side wall of faecal sludge chamber. There are health risks to the workers and there is no control over where the disposal of the excavated material although this unregulated practice is relatively cheap to keep the pit latrine operational.

A number of alternative technologies have been developed in the last two decades, to overcome the high haulage costs and access problems associated with mechanized pit emptying services in slums. The Vacutug MK1, Vacutug MK 2 and the MAPET (Manual Pit Emptying Technology) have been tested in Africa and Asia (Table 2.6). They are combinations of a tank with a capacity range of 200 to 500 litre capacity (Thye et al., 2011), a small manual or motorized pump for extracting faecal sludge connected to a flexible horse pipe and wheels suited for manoeuvring in congested area. These technologies require less skilled labour and have low local operation and maintenance cost. Their limitations include haulage distances in excess of 0.5 km to the treatment unit, weak pit latrine substructure in the case of pumping, depth of less than 2m and inability to remove dry sludge and solid particles like stones and wood from pits (Boot and Scott, 2008; Harvey, 2007).

Secondary treatment of the unstabilised sludge after frequent emptying of pit latrines in urban slums is required. Sludge settling/thickening tanks and sludge drying beds are used for treatment of sludge in developing countries before land application or re-use (Cofie et al., 2006; Ingallinella et al., 2002; Koné and Strauss, 2004). They are a low cost option for hot climate suitable for decentralized application at locations distant from settlements where land is available to avoid odour and flies (Tilley et al., 2008). The dewatered and stabilized sludge can be used for horticulture and land application in urban areas. Pit latrines serve as an interface in the sanitation system and can only be sustainable if there is collection, treatment and reuse or safe disposal of sludge. Institutional support is required for continuity of pit latrine emptying services in addition to technology and social and economic aspects (Thye et al., 2011; Franceys, 2008).

Table 2.6:   A comparison of Vacutug and MAPET technologies for faecal sludge emptying (Issaias, 2006; NWSC, 2008; Thye et al., 2011)

Technology	Vacutug MK1	Vacutug MK2	MAPET
First application	Kibera slum, Nairobi, Kenya (1995).	Dhaka, Bangladesh in (1999).	Dar es Salaam, Tanzania (1989).
Components of system	500 litre tank, vacuum pump powered by small petrol engine with hose and hand cart.	As Vacutug, in two sections with smaller (200 L) tank in one section and a remote 1900 L collection tank	200 litre tank, vacuum created by hand pump mechanism, hose and hand cart.
Access width (m)	1.5	> 1	> 1
Depth (m)	≤ 2	≤ 2	≤ 2
Applying conditions	Areas with high density of population using pit latrines. Access corridor of 1.5 m width is adequate. Availability of fuel for the motorised system.	Areas with poor accessibility and narrow corridors between housing units. Petrol needs to be relatively cheap and easily available.	Areas where fuel for motorised systems is not affordable and where sludge can be pumped easily.
Current Status	Still in use in Kibera. Largely superseded elsewhere by the MK 2.	In use in more than 10 cities in developing countries.	Little mention in recent literature.

2.3.3.2   Biogas toilets

The anaerobic digestion of faecal matter has been applied in India and Kibera slum (Kenya) for economic gain from the produced biogas as energy source. An innovative Sulabh biogas toilet (Figure 2.4) that utilizes anaerobic digestion to produce biogas from human excreta was developed in India and has been recognized for contributing towards the sanitation MDG target (Pathak, 2009). The facility consists of a toilet connected to a digester and biogas is stored under the fixed dome by hydraulic displacement of the digesting slurry inside the digester (Sulabh International, 1999). This technology has been implemented in slums in India at households for excreta disposal and for use in slums as public toilets linked to a biogas plant or effluent treatment systems for nutrient recycling (Jha, 2005).

A   B

Figure 2.4:   **Sulabh biogas toilet system in India: A) Public toilet linked to a digester B) Effluent treatment part of the system (Pathak, 2009)**

A similar innovative storeyed facility with a restaurant, meetings hall and bio-toilets (Figure 2.5) was implemented in Kibera slum financed by Umande Trust (Local NGO) and managed by TOSHA CBO (Schouten and Mathenge, 2010). It has potential for co-digestion of excreta and organic waste from the restaurant to increase biogas production, which is used for cooking and water heating. It has provided economic and health benefits such as employment and energy recovery from waste and pollution control. Small-scale biogas systems for the treatment of faeces and kitchen waste have also been implemented in Kochi, South India (Estoppey, 2010).

Figure 2.5: Biogas toilet in Kibera slum, Kenya

Biogas toilets need to be implemented within biogas sanitation systems that address the transportation, storage, treatment, and reuse or disposal of the end products (Mang and Li, 2010). Implementation of biogas toilets requires skilled personnel at local level for process control. The occurrence of pathogens in the digestate and the investments cost that are still not affordable by the majority in the targeted communities remain unexamined issues despite the good social acceptance of biogas toilets. The results of application of this technology in slums look promising although its effectiveness in the long term remains to be investigated.

2.3.3.3 Sulabh flush compost toilet

The sulabh flush compost toilet is one of the sanitation systems that have been used in urban slums (Sulabh International, 1999; WaterAid India, 2006). It is a two pit pour flush toilet made of brick work and concrete and has been developed in India for safe disposal of human excreta (Pathak, 2009; Sulabh International, 1999). Sulabh toilets have sludge holding pits that are not water proof to allow for exfiltration and should not be built in areas with a high water table and less than 30 m upstream of a drinking water source. Sludge stored in a pit for a period of 2 to 3 years undergoes composting after which it is safely removed as a dry manure or soil conditioner with low levels of pathogens that may not pose a risk during handling. This type of sanitation facility can also be used in urban slums located in areas with low water table since it is simple and inexpensive in terms of construction and operation and uses less water.

2.3.3.4 Community based sanitation facilities

Community based sanitation in the form of communal sanitation blocks provide a good and affordable alternative for excreta disposal (Carter et al., 1999; Mara, 2003). The blocks should be maintained by caretakers who collect a user fee for the operation and maintenance although under the MDG criteria, public facilities and shared facilities by two or more households are considered unimproved by WHO. Sulabh technologies, SPARC (non-profit construction company in India) community built toilets and Kibera biogas blocks are examples of best practices in providing sanitation for the urban poor. They have common attributes: people are willing to pay; there is a high community involvement and they fit in the local circumstances. The costs depend on raw materials and labour that vary in different countries. Community based sanitation has been successfully implemented in slums in India, Lao PDR and Vietnam (Pathak, 2009, UN-Habitat, 2006). Communal sanitation has also been found suitable in Kibera slum in Kenya (Schouten and Mathenge, 2010). Community based approaches have been tested in urban slums and have shown to have the potential to improve the sustainability of sanitation interventions (Lüthi et al., 2009). Community toilets can also be a tool for bringing behavioural changes in slums with combined with sensitisation and awareness campaigns.

2.3.3.5   Peepoo bag

Peepoo bags were developed as a result of the unhygienic practices of disposing human waste in polythene bags commonly referred to as "flying toilets" because they are used and disposed of indiscriminately. The peepoo bag has outer dimensions of 15x40 cm^2 with a gauze liner (Figure 2.6) and is coated on the inside with a film of urea (4g) as the sanitising agent (Ondieki and Mbegera, 2009; Vinnerås et al., 2009). The concept is that by the contact between urea and faeces, breakdown of urea into ammonium and carbonate raises the pH and triggers hygienisation resulting in elimination of viruses, bacteria and parasites over time, depending on the environmental conditions (Vinnerås et al., 2009).

The peepoo bags are biodegradable and need to be disposed of at a storage place to allow for hygienisation and thereafter reuse as a fertiliser (Ondieki and Mbegera, 2009). It is important to emphasize that their use must be combined with other hygiene practices such as safe disposal and washing of hands in order to achieve a total sanitation solution. Community sensitization is important to prevent a health hazard to the public, especially contact of children with used Peepoo bags. For large scale application of this technology in slums, user satisfaction, acceptance and affordability should be considered. A collection system for the used peepoo bags and the marketing of the sanitised product are required for this innovation to be sustainable as a substitute for "flying toilets" in urban slums. Peepoo bags could be used in urban slums although they are not considered as basic sanitation under the MDGs.

(A)   (B)

Figure 2.6:   (A) The peepoo bag with the outer bag folded down; holding the bag with hand covered by inner foil; and the used bag (Vinnerås et al., 2009) and (B) Use of peepoo bag in slums (Kenya)

2.3.3.6 Innovations in faecal sludge collection and treatment

There have been innovations in recent years in the field of collection and treatment of faecal sludge. They were based on accepted sanitation technologies in use such as pit latrines, bucket latrines or "flying toilets". These innovations did not aim at introducing a new type of toilet but rather to provide easier collection, better treatment and more dignity, and to minimise public health hazards and environmental pollution. This was achieved by retaining the excreta either mixed or separated. Examples are the peepoo bag and the Ghanasan, which is an alternative to the bucket latrine. The Ghanasan has a removable sealed compartment underneath for collection and transportation of excreta for treatment. A combination of a sealed container and provision of a good service could make the system sustainable.

A treatment step is incorporated in some technologies to reduce transportation costs and to produce a product that is locally useful such as biogas from biogas toilets. These examples show that innovation can be simple, minimise contact with excreta and provide for the collection and treatment of faecal sludge. Exclusion of non-biodegradable solid waste from the contents of collection chambers or containers is a challenge for some technologies like biogas toilets and UDDT. In addition, solid waste reduces the life span of pit latrines (Still et al., 2010). This can be overcome by either providing an effective solid waste management system or having non-shared sanitary facilities which is unlikely in slums.

2.3.4 Collection and treatment of urine

2.3.4.1 Urine diversion dehydrating toilet

A urine diversion dehydrating toilet (UDDT) as an excreta disposal option separates urine and faecal matter. They are currently used in peri-urban and urban areas of East Africa, South Africa, South America and Asia (Rieck and Von Münch, 2009). It can be constructed and repaired with locally available materials even in rocky areas and does not require water. A UDDT is advantageous in areas that are prone to floods when faecal matter is stored in a water proof chamber and urine is collected separately. Public health risks associated with spillage of mixed waste streams in the environment are minimised (Esrey et al., 1998). A provision for "washers" to divert wastewater from anal cleansing to a separate container can be inserted. A small area of land is required for construction of a UDDT while the capital costs depend on the materials used. Blockages of the urine diverting pipe in a UDDT can be caused by foreign particles and precipitation of struvite ($MgNH_4PO_4$) and calcium phosphates ($Ca_{10}(PO_4)_6(OH)_2$) when the pH increases (Kvarnström et al., 2006). This can be overcome by use of pipes with a large diameter and elimination of sharp bends in the urine diversion conduits.

Manual removal of the waste products and safe disposal or reuse of the urine and faecal matter are necessary for the sustainability of a UDDT as a sanitation option in slums. Inactivation of pathogens is achieved by adding ash or lime to faecal matter to reduce odour and increase pH (Kvarnström et al., 2006). Other factors that influence pathogen die-off in the faecal matter include temperature, water content in the material and exposure time (Cools et al., 2001; Nakagawa et al., 2006; Vinneras et al., 2008; Niwagaba et al., 2009; Martens and Böhm, 2009). Urine can be safely used after storage in containers for 6 months at 20 ^{0}C for unrestricted use (Höglund et al., 2002; Vinnerås et al., 2008).

The sanitised dry faecal matter is stored in sacks or containers in the back yard ready for use as a fertiliser while sanitised urine stored in plastic containers can be directly applied to soil, compost and crops (Kvarnström et al., 2006; Höglund et al., 2002). Urine could be used as an alternative to chemical fertilisers although research on the fate of pharmaceutical residues is required (Karak and Bhattacharyya, 2010). Collection and transportation of these end products in slums is by use of carts and small trucks depending on accessibility. UDDTs are suitable facilities to use with an aim of closing the nutrient loop through reuse of the sanitised urine and excreta as a soil conditioner.

2.3.4.2 Struvite precipitation

Struvite precipitation is an innovative technology for nutrient recovery from liquid waste streams, such as human urine. Recovery of nutrients from human urine has attracted a lot of attention in an effort towards closing the loop between urine generation and reuse of nutrients as fertilisers. Urine constitutes the major fraction of nitrogen, phosphorus and potassium in domestic wastewater which are essential components of plant fertilizers (Meinzinger and Oldenburg, 2008). Several studies have been carried out on nutrient recovery from urine using activated carbon as adsorbent, struvite precipitation and mineral adsorption (Lind et al., 2000; Ronteltap et al., 2007). Struvite and activated carbon act as slow release fertilizers in terms of nitrogen plant availability (Ganrot et al., 2007; Udert et al., 2003).

Urine-separation toilets are considered a strategic route to maximize urine nutrient recovery by transforming it into a solid mineral (Lind et al., 2000). The concept of struvite precipitation can also be applied to the liquid phase separated from biosolids after anaerobic digestion. Though this technology seems promising, its application in developing countries in urban slums is rather unlikely in the near future due to the required chemical dosage control and high costs.

2.3.5 Collection and treatment of solid waste and faecal sludge for resource recovery

2.3.5.1 Anaerobic co-digestion

Anaerobic co-digestion can be applied in slums as part of the sustainable solution to generate biogas and tackle public health and environmental problems associated with poor solid waste and faecal sludge management (Deublein and Steinhauser, 2008; Jingura and Matengaifa, 2009; Lettinga et al., 2001; Lettinga, 2008). Adapting and implementing digesters for anaerobic co-digestion of organic solid waste and excreta in urban slums is driven by: increased yield of biogas and use of the end product as a fertiliser although its post treatment is required for further destruction of pathogens (Appels et al., 2008; Mata-Alvarez et al., 2000; Neves et al., 2008; von Münch, 2008). Digesters have been implemented in slum areas in Kenya, Uganda, and Zimbabwe at both household and community level and on a big scale in India using cow dung mixed with organic solid waste and excreta as the feed material (Jingura and Matengaifa, 2009; Kapdi et al., 2005; Müller, 2007; Viswanath et al., 1992).

The types of anaerobic digesters that can be implemented as part of the onsite and decentralized sanitation systems approach in slums include continuous wet and dry fixed domes, continuous plug flow and batch dry fixed domes (Jha et al., 2011; Müller, 2007; von Münch, 2008). Their operational conditions in slums are temperature range of 20-35 °C for mesophilic digestion, total solids content below 20% for wet systems and 25% to 50% for dry systems, and the hydraulic retention of 15 to 30 days (de Mes et al., 2003; Müller, 2007). The volatile solids loading rate range is 1.6 to 4.8 kg per m^3 of active digester volume per day for complete mix mesophilic anaerobic digestion. The biogas yield ranges from 0.25 to 0.95 m^3 per kg volatile solids depending on the feed stock (Appels et al., 2008; Jha et al., 2011; WEF and ASCE, 1998). Pre-treatment of mixed biodegradable waste to reduce their size improves digester performance in terms of biogas yield (Kujawa-Roeleveld and Zeeman, 2006; Mata-Alvarez et al., 2000). Exclusion of non-biodegradable substances for anal cleansing is important when applying anaerobic digestion as the treatment process (Schönning and Stenström, 2004). The solid waste can be collected in slums using containers and bins at household level and transported to digesters using hand carts and small trucks supported by institutional arrangements with private sector involvement.

The sustainable application of anaerobic digestion in urban slum is a challenge. Post treatment of the digestate after mesophilic anaerobic digestion to eliminate pathogens for reuse using the multi-barrier approach is required. This is especially true in developing countries where energy requirements for operation of digesters under thermophilic conditions can be costly. The current trend in research focuses on development of cost-effective technologies for effluent post treatment and recovery of mineral concentrates to make anaerobic co-digestion a core technology for treatment of mixed solid waste, manure and human excreta (Chen et al., 2008; Holm-

Nielsen et al., 2009; Jha et al., 2011; Jingura and Matengaifa, 2009; Reith et al., 2003). Other challenges of operating digesters in slums include the highly skilled labour and seeding required. The implementation of anaerobic digesters in slums like other development projects has disadvantages that include potential public health risks and negative environment impacts. The pathogen content of the feed stock and the digestate poses risks to human health. There may be risks of fire and explosion and eventual ground water contamination from leachate.

2.3.5.2 Composting

Composting is one of the processes that can be used to treat excreta alone or mixed with other organic matter to achieve an optimum carbon to nitrogen ratio (Lopez Zavala et al., 2004; Niwagaba et al., 2009, Strauss et al., 2003). The faecal sludge from slums can be collected and transported by the Vacutug and MAPET while the solid waste can be transported by hand carts and small trucks to the composting site. A rapid inactivation of pathogens occurs during composting due to different factors such as antibiosis, pH-shift, redox-potential, antagonism (hostility among bacterial groups), nutrient deficiencies and exothermic metabolism. The main factors influencing the inactivation of pathogens are temperature, water content, exposure time, and carbon to nitrogen (C:N) ratio of 15:1 to 30:1 (Huang et al., 2004; Lopez Zavala et al., 2004; Martens and Böhm, 2009; Niwagaba et al., 2009; Vinnerås, 2007; Vinnerås et al., 2008). Addition of green grass clippings, vegetable scraps, straw, husks and wood shavings that are found in slums can increase the C:N ratio, provide oxygen to the pile and help to achieve rapid and complete decomposition. The product is hygienically safe and rich in humus carbon, fibrous material, nitrogen, phosphorus and potassium (Cools et al., 2001; Lopez Zavala et al., 2004; Martens and Böhm, 2009; Niwagaba et al., 2009; Vinnerås et al., 2008).

Co-composting of excreta, manure and other organic materials provides an opportunity to treat excreta from urine diversion dehydration toilets, and anaerobically digested sludge in the mesophilic range (Huang et al., 2004; Vinnerås, 2007). Excluding urine (such as in urine diverting toilets) lowers the amount of nitrogen in relation to carbon (Vinnerås et al., 2008). Co-composting of excreta with organic waste has potential for application in urban slums to recover nutrients in the form of a soil conditioner. The limitations to application of composting in slums are: lack of space, capital cost for the facility infrastructure and aeration requirements, and flooding effects that may cause environmental pollution and public health concerns in low-lying areas.

2.3.5.3 Pyrolysis

Pyrolysis has potential to be applied for resource recovery at municipal sewage treatment plants where cesspool emptiers discharge sludge from slums. In developing countries where energy costs are high, there is demand for development of cost effective methods for separation of digested sludge into liquid and solid phases for subsequent processing. The liquid concentrate can be recovered as nitrate-phosphate fertiliser (Aiyuk et al., 2004) and the solid phase subjected to pyrolysis for biochar production at a high temperature of 400 °C to 700 °C (Bruun et al., 2008; Demirbas, 2004; Gaunt and Lehmann, 2008; Gaskin et al., 2008; Lehmann et al., 2006; Winsley, 2007).

A special type of pyrolysis is called hydrothermal carbonization (HTC), which is defined as the conversion of biomass into coal under wet conditions and low temperature (around 200 °C). Coalification of biomass is a natural chemical process, but takes place on the time scale of some hundred (peat) to millions (black coal) of years. Recently, it was discovered that the presence of iron can effectively accelerate HTC, which shortens the process to only a few hours. Since then, HTC has been demonstrated for a vast variety of wet organic wastes in simple and inexpensive experimental set-ups (Titirici et al., 2007). Producing biochar from faecal sludge in densely populated areas can have various beneficial effects. These not only include better hygiene conditions but also an increase in the local supply of cooking fuels (Paraknowitsch et al., 2009), use as a soil conditioner for improved agricultural yields (Gaunt and Lehmann, 2008; Steinbeiss et al., 2009) and in water and wastewater treatment applications (Figure 2.7). The challenges to application of this technology in slums include: the high cost of energy, the risk to the public from explosions due to high temperature and pressure, and the market for the biochar. There is also uncertainty of the net emissions of green house gases linked to the biochar cycle due to limited experimental data (Lehmann et al., 2006; Ogawa et al., 2006; Reijnders, 2009). Decentralised units need to be tested at pilot scale to assess the feasibility of this technology as a faecal sludge treatment step of a sanitation system.

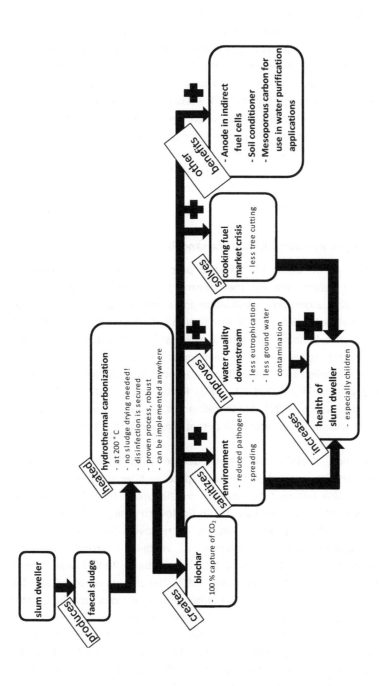

Figure 2.7: Benefits of the conversion of faecal sludge into biochar or charcoal via the hydrothermal carbonization process and its subsequent positive effects (marked with a '+'sign)

2.3.6   Collection and treatment of grey water

2.3.6.1   Collection of grey water

Grey water can be collected and transported by pipes where simplified sewerage exists or it can be collected in buckets at household level and applied to a nearby soil or sand filter for treatment. Current grey water management practices in slums include use of buckets to discharge grey water into poor constructed open storm water drains and open space resulting in ponding and to a lesser extent reuse in household gardens (Parkinson and Tayler, 2003; Kulabako et al., 2009; Mungai, 2008). Agricultural reuse is attributed to knowledge of grey water nutrient content by some inhabitants while pouring it in the open space is partly because lack of awareness of the health risks to children who use the open space as playing grounds (Carden et al., 2007; Imhof and Muhlemann, 2005). Community sensitization is important to increase both awareness of health risks from contact with untreated grey water and collection efficiency using buckets before treatment.

2.3.6.2   Treatment of grey water with soil and sand filters

Soil and sand filter systems utilise the filtration process for grey water treatment. Filtration involves pollutant removal mechanisms that include: biodegradation, straining, sedimentation, adsorption, nitrification and denitrification. Biodegradation is the primary removal mechanism of organic matter in both the solid and liquid phase performed by the active biomass attached to solid surfaces (Campos et al., 2002; Rauch et al., 2005). The highest biomass activity occurs in the first 10-30 cm of the infiltration surface as applied to vertical filtration in both natural and artificial filter systems (Maksimović and Tejada-Guibert, 2001; Rauch et al., 2005; Ridderstolpe, 2007; Von Felde and Kunst, 1997). Biodegradation therefore, contributes to the sustainability of soil and sand filters and their potential application in slums.

Physical and mechanical straining can also be utilised using locally available filter media to remove particles larger than the pore space. In filter systems, particulate Kjeldahl nitrogen is also removed by filtration in the upper filter layer and the dissolved part by adsorption onto media particles (Idelovitch et al., 2003; Metcalf and Eddy, 2003; Tchobanoglous, 1998). Ammonia is removed by adsorption and nitrification-denitrification while phosphorus is removed through chemical precipitation and adsorption (Bayley et al., 2003; Carr and Baird, 2000; Panuvatvanich et al., 2009; Sedlak, 1991; Von Felde and Kunst, 1997). The effluent is not suitable for potable use because it still contains bacteria although their removal may be greater than 2 log (De Leon et al., 1986; Hijnen et al., 2004). In addition, viruses tend to survive longer due to their small size and ability to adapt to different conditions through active and inactive stages (Nema et al., 2004). Soil and sand filters can be applied in slums because they are robust systems made of locally available materials and can be implemented at household level.

2.3.6.3 Application of grey water fed soil and sand filter systems in slums

Filtration systems have been applied in slums for grey water treatment. Mulch towers are an example of a low cost technology for grey water treatment being used in Southern African (Zuma et al., 2009). Grey water towers (Figure 2.8) have also been used in East Africa for grey water treatment and nutrient recovery (Kulabako et al., 2009) although the risk to humans is high for consumption of raw vegetables irrigated with grey water (Siobhan et al., 2006). There is also a potential risk to the slum dwellers through leakage of micro-pollutants to ground water in the long term and also a potential problem for the environment (Larsen and Maurer, 2011; Joss, 2006). These treatment units require 1 to 3 m^2 of land, well graded media with coefficient of uniformity less than 5 and an oil and grease trap. Addition of macro nutrients (N, P) and trace metals (Zn, Al) to nutrient-deficient grey water was found to increase grey water treatment efficiency (Jefferson et al., 2001). In urban slums, a balanced C: N may be achieved by mixing grey water from the kitchen with other grey water types from laundry and bathroom. Other types of similar filter systems that have been used for grey water treatment include the horizontal flow trench system, the crate soakaway and intermittent sand filters.

(A)   (B)

Figure 2.8: (A) Completed grey water tower and (B) Onions planted on the sides and tomatoes planted on top of a grey water tower to remove nutrients (Kulabako, 2009)

The varying composition of grey water (Table 2.4) implies a need to apply low- cost and robust systems for grey water treatment. Soil and sand filter units can be applied in urban slums at household and institutional levels since they are made of locally available materials, affordable and require limited space and less skilled personnel. The challenges in applying this technology in slums include variability of the quantity and quality of grey water and the intermittent flow in the filters. The acceptability by the users and the treatment efficiency in relation to the quantity of grey water generated in slums are also important for the sustainability of soil and sand filters in a sanitation system.

2.4 Conclusion

Pit latrines are the dominant type of excreta disposal facilities in urban slums. However, it is difficult to operate them on a sustainable basis because they pollute drinking water sources and faecal sludge management is inadequate. The increasing amount of uncollected mixed solid waste and grey water being discharged into the environment without treatment is also a constraint for sustainability of urban slum sanitation. Sustainability requires that institutional arrangements and collection of the waste, treatment, reuse and safe disposal complement each other. It is affected by interlinked factors including low household incomes, lack of financial incentives for service providers, limited accessibility, lack of ownership and municipal recognition. A number of technologies could be applied in slums subject to the feasibility criteria in different practical situations. Simplified sewerage may be implemented in informal settlements with piped water supply and space for the wastewater treatment facility when recovery of nutrients is not an objective. Anaerobic co-digestion of excreta and organic solid waste may be feasible in urban slums for decentralized treatment of sludge and solid waste to produce biogas, provided the digestate is sanitised for safe reuse or disposal and there is skilled labour for process control. UDDT is an attractive low cost technology option for urban slums that separates human excreta to minimise health risks and for nutrient recycling depending on the availability of land for construction and of a market for the end product.

Grey water management in urban slums has not been given priority, despite being the largest wastewater stream and polluted with pathogens, nutrients and micro-pollutants. There is potential to increase pollutant removal efficiencies from grey water before its disposal or reuse using optimised soil and sand filters that occupy less space. These sanitation technologies need to be tested at pilot scale to evaluate their performance in slums as part of a sanitation system prior to up scaling for wide application. Acceptability by the users and existence of an enabling institutional framework improve chances of technology success. A quantitative risk assessment approach for the sanitation interventions could be used as a basis to maximize public health benefits, achieve the sanitation MDG target and minimise environmental pollution in slums.

References

Aiyuk, S., Amoako, J., Raskin, L., van Haandel, A., Verstraete, W., 2004. Removal of carbon and nutrients from domestic wastewater using a low investment, integrated treatment concept. Water Research 38, 3031-3042.

Appels, L., Baeyens, J., Degréve, J., Dewil, R., 2008. Principles and potential of the anaerobic digestion of waste-activated sludge. Progress in Energy and Combustion Science 34, 755–781.

Ashbolt, N.J., 2004. Microbial contamination of drinking water and disease outcomes in developing regions. Toxicology 198, 229–238.

Avvannavar, S.M., Mani, M., 2008. A conceptual model of people's approach to sanitation. Science of the Total Environment 390, 1-12.

Bakalian, A., Wright, A., Otis, R., Neto J.A., 1994. Simplified sewerage guidelines. UNDP – World Bank Water and Sanitation Program, World Bank, Washington DC.

Bayley, M.L., Davison, L., Headley, T.R., 2003. Nitrogen removal from domestic effluent using subsurface flow constructed wetlands: influence of depth, hydraulic residence time and pre-nitrification. Water Science & Technology 48(5), 175–182.

Bhatia, M.S., Gurnani, P.T., 1996. Urban waste management privatization. Reaching the unreached: Challenge for 21st century. 22nd WEDC conference, New Delhi, India.

Boot, N.L.D., Scott, R.E., 2008. Faecal sludge management in Accra, Ghana: Strengthening links in the chain. 33rd WEDC International Conference, Accra, Ghana.

Bracken, P., Kvarnström, E., Ysunza, A., Kärrman, E., Finnson, A., Saywell, D., 2005. Making sustainable choices – the development and use of sustainability oriented criteria in sanitary decision making. Third International Conference on Ecological Sanitation, Durban, South Africa.

Bracken, P., Wachtler, A., Panesar, A.R., Lange, J., 2007. The road not taken: how traditional excreta and greywater management may point the way to a sustainable future. Water Science & Technology, Water Supply 7(1), 219-227.

Burra, S., Patel, S., Kerr, T., 2003. Community-designed, built and managed toilet blocks in Indian cities. Environment&Urbanization 15(2), 11-32.

Cairncross, S., 1992. Sanitation and water supply: practical lessons from the decade. UNDP – World Bank Water and Sanitation Program, The International Bank for Reconstruction and Development/The World Bank, Washington DC.

Carden, K., Armitage, N., Winter, K., Sichone, O., Rivett, U., Kahonde, J., 2007. The use and disposal of grey water in the non-sewered areas of South Africa: Part 1- Quantifying the grey water generated and assessing its quality. Water SA 33, 425-432.

Carter, R.C., Tyrrel, S.F., Howsam, P., 1999. Impact andsustainability of community water supply and sanitation programmes in developing countries. Journal of Chartered Institution of Water and Environmental Management 13, 292-296.

Chaggu, E.J., Mashauri, A., Van Buuren, J., Sanders, W., Lettinga, J., 2002. PROFILE: Excreta Disposal in Dar es Salaam. Environment Management 30(5), 609-620.

Chanakya, H.N., Sharma, I., Ramachandra, T.V., 2009. Micro-scale anaerobic digestion of point- source components of organic fraction of municipal solid waste. Waste Management 29, 1306-1312.

Chen, Y., Cheng, J.J., Creamer, K.S., 2008. Inhibition of anaerobic digestion process: A review. Bioresource Technology 99, 4044-4064.

Christova-Boal, D., Eden, R.E., McFarlane, S., 1996. An investigation into grey water reuse for urban residential properties. Desalination 106(1-3), 391-397.

Clara, M., Kreuzinger, N., Strenn, B., Gans, O., Kroiss, H., 2005. The solids retention time - a suitable design parameter to evaluate the capacity of wastewater treatment plants to remove micropollutants. Water Research 39(1), 97-106.

Cofie, O.O., Agbott, S., Strauss, M., Essek, H., Montanger, A., Awua, E., Kone, D., 2006. Solid–liquid separation of faecal sludge using drying beds in Ghana: Implications for nutrient recycling in urban agriculture. Water Research 40, 75- 82.

Cools, D., Merckx, R., Vlassak, K., Verhaegen, J., 2001. Survival of E.coli and Enterococcus spp. Derived from pig slurry in soils of different texture. Applied Soil Ecology 17, 53-62.

Crites, R.W., Tchobanoglous, G., 1998. Small and Decentralized Wastewater Management Systems. McGraw-Hill Companies, California, ISBN 0-07-289087-8 USA.

De Leon, R., Singh, S.N., Rose, J.B., Mullinax, R.L., Musial, C.E., Kutz, S.M., Sinclair, N.A., Gerba, C.P., 1986. Microorganism removal from wastewater by rapid mixed media filtration. Water Research 20 (5), 583-587.

de Mes, T.Z.D., Stams, A.J.M., Zeeman. G., 2003. Methane production by anaerobic digestion of wastewater and solid wastes. In: Reith JH, Wijffels RH, Barten, H, editors. Biomethane and Biohydrogen. Status and perspectives of biological methane and hydrogen production, p. 58-94.

Demirbas, A., 2004. Effects of temperature and particle size on bio-char yield from pyrolysis of agricultural residues. Journal of Analytical and Applied Pyrolysis 72(2), 243-248.

Dzwairo, B., Hoko, Z., Love, D., Guzha, E., 2006. Assessment of the impacts of pit latrines on groundwater quality in rural areas: A case study from Marondera district, Zimbabwe. Physics and Chemistry of the Earth, Parts A/B/C 31(15-16), 779-788.

Ellis, J.B., Hvitved-Jacobsen, T., 1996. Urban drainage impacts on receiving waters. Journal of Hydraulic Research 34(6), 771-783.

Elmitwalli, T., Mahmoud, N., Soons, J., Zeeman, G., 2003. Characteristics of grey water: Polderdrift, the Netherlands. Proceedings of 2nd International Symposium on Ecological Sanitation. Luebeck, Germany.

Elmitwalli, T., Otterpohl, R., 2007. Anaerobic biodegradability and treatment of grey water in upflow anaerobic sludge blanket (UASB) reactor. Water Research 41(6), 1379-1387.

Eriksson, E., Andersen, H.R., Madsen, T.S., Ledin, A., 2009. Grey water pollution variability and loadings. Ecological Engineering 35, 661-669.

Eriksson, E., Aufarth, K., Henze, M., Ledin, A., 2002. Characteristics of grey wastewater. Urban Water 4, 85-104.

Esrey, S.A., Gough, J., Rapaport, D., Sawyer, R., Simpson-Hebert, M., Vargas, J., Winblad, U., 1998. Ecological Sanitation. SIDA, Stockholm, Sweden. http://www.ecosanres.org/pdf_files/Ecological_Sanitation.pdf [Accessed on 21st October, 2010].

Estoppey, N., 2010. Evaluation of small-scale biogas systems for the treatment of faeces and kitchen waste. Eawag-SANDEC publications.

Fang, A., 1999. On-site sanitation: an international review of World Bank experience. UNDP- World Bank Water and Sanitation Program-South Asia, Washington DC, USA.

Feachem, R.G., Bradley, D., Garelick, H., Mara, D., 1983. Sanitation and Disease: health espects of excreta and wastewater management, John Wiley and sons, Chichester, UK.

Foppen, J.W.A., 2002. Impact of hugh-strength wastewater infiltration on groundwater quality and drinking water supply: the case of Sana,a, Yemen. Journal of Hydrology 263, 198-216.

Franceys, R., 2008. 'Running Institutions or Walking the Talk', Stockholm Water Week. http://www.worldwaterweek.org/documents/WWW_PDF/2008/thursday/K16_17/Richard_Franceys_Sanitation.pdf. (Accesses on Oct. 10, 2010).

Ganrot, Z., Dave, G., Nilsson, E., Li, B., 2007. Plant availability of nutrients recovered as solids from human urine tested in climate chamber on Tritium aestivum L. Bioresource Technology 98, 3122-3129.

Gaskin, J.W., Steiner, C., Harris, K., Das, K.C., Bibens, B., 2008. Effect of low-temperature pyrolysis conditions on biochar for agricultural use. Transactions of the American Society of Agricultural and Biological Engineers 51(6), 2061-2069.

Gaunt, J.L., Lehmann, J., 2008. Energy Balance and Emissions Associated with Biochar Sequestration and Pyrolysis Bioenergy Production. Environmental Science and Technology 42(11), 4152-4158.

Genser, B., Strina, A., dos Santos, L.A., Teles, C.A., Prado, M.S., Cairncross, S., Barreto, M.L., 2008. Impact of a city-wide sanitation intervention in a large urban centre on social, environmental and behavioural determinants of childhood diarrhoea: analysis of two cohort studies. International Journal of Epidemiology 37(4), 831-840.

Gerba, C.P., Straub, T.M., Rose, J.B., Karpiscak, M.M., Foster, K.E., Brittain, R.G., 1995. Water quality of grey water treatment system. Water Research 31(1), 109-116.

GTZ (Deutsche Gesellschaft für Technische Zusammenarbeit), 2002. Ecosan-recycling beats disposal. Eschborn, Germany.

Hargelius, K., Holmstrand, O., Karlsson, L., 1995. Hushallsspillvatten Framtagande av nya schablonvärden för BDT-vatten. In Vad innehaller avlopp fran hushall?

Näring och metaller i urin och fekalier samt i disk-, tvätt, bad- and duschvatten. Stockholm: Swedish EPA (in swedish).

Harvey, P.A., 2007. Excreta Disposal in Emergencies: A field manual. Loughborough University, Water Engineering and Development Centre.

Heinss, U., Larmie, S.A., Strauss, M., 1994. Sedimentation Tank Sludge Accumulation Study. EAWAG/SANDEC publications.

Heinss, U., Larmie, S.A., Strauss, M., 1998. Solids Separation and Pond Systems for the Treatment of Septage and Public Toilet Sludges in Tropical Climate – Lessons Learnt and Recommendations for Preliminary Design. EAWAG/SANDEC Report No. 05/98.

Hijnen, W.A., Schijven, J.F., Bonné, P., Visser, A., Medema, G.J., 2004. Elimination of viruses, bacteria and protozoan oocysts by slow sand filtration. Water Science & Technology 50(1), 147-54.

Höglund, C., Stenström, T.A., Ashbolt, N., 2002. 'Microbial risk assessment of source-separated urine used in agriculture.' Waste Management Research 20(3), 150-161.

Holm-Nielsen, B., Al Seadi, T., Oleskowicz-Popiel, P., 2009. The future of anaerobic digestion and biogas utilization. Bioresource Technology 100(22), 5478-5484

Howard, G., Pedley, S., Barret, M., Nalubega, M., Johal, K., 2003. Risk factors contributing to microbiological contamination of shallow groundwater in Kampala, Uganda. Water Research 37, 3421–9.

Huang, G.F., Wong, J.W.C., Wu, Q.T., Nagar., 2004. Effect of C/N on composting of pig manure with saw dust. Waste management 24, 805-813.

Imhof, B., Muhlemann, J., 2005. Greywater treatment on household level in developing countries-A state of the art review. Department of Environmental Sciences at the Swiss Federal Institute of Technology (ETH), Zürich, Switzerland.

Ingallinella, M., Sanguinetti, G., Koottatep, T., Montangero, A., Strauss, M., 2002. The challenge of faecal sludge management in urban areas – strategies, regulations and treatment options. Water Science and Technology 46(10), 285-294.

Issaias, I., 2006. Vacutug. Faecal sludge Management Symposium 9th-11th May, UN-Habitant; http://siteresources.worldbank.org/EXTWAT/Resources/4602122-1215104787836/FSM_UNHabitat.pdf [Accessed on 1st February, 2011].

Isunju, J.B., Schwartz, K., Schouten, M.A., Johnson, W.P., van Dijk, M.P., 2011. Socio-economic aspects of improved sanitation in slums: A review. Public Health 125, 368-376.

Jefferson, B., Burgess, J.E., Pichon, A., Harkness, J., Judd, S.J., 2001. Nutrient addition to enhance biological treatment of grey water, Water Research 35(1), 2702-2710.

Jefferson, B., Laine, A., Parsons, S., Stephenson, T., Judd, S., 1999. Technologies for domestic wastewater recycling. Urban water 1(4), 285-292.

Jenkins, M.W., Sugden, S., 2006. Rethinking sanitation. Lessons and innovation for sustainability and success in the New Millennium. UNDP - sanitation thematic paper.

Jha, A.K., Li, J., Nies, L., Zhang, L., 2011. Research advances in dry anaerobic digestion process of solid organic wastes. African Journal of Biotechnology 10(65), 14242-14253.

Jha, P.K., 2005. Recycling and reuse of human excreta from public toilets through biogas generation to improve sanitation, community health and Environment. Sulabh International Academy of Environmental sanitation.

Jha, P.K., 2003. Health and social benefits from improving community hygiene and sanitation: an Indian experience. International Journal of Health Research Suppl 1, S133-40.

Jingura, R.M., Matengaifa, R., 2009. Optimisation of biogas production by anaerobic digestion for sustainable energy development in Zimbabwe. Renewable and Sustainable Energy Reviews 13, 1116-1120.

Joss, A., Zabczynski, S., Göbel, A., Hoffmann, B., Löffler, D., McArdell, C.S., Ternes, T.A., Thomsen, A., Siegrist, H., 2006. Biological degradation of pharmaceuticals in municipal wastewater treatment: Proposing a classification scheme. Water Research 40(8), 1686-1696.

Joyce, J., Granit, J., Frot, E., Hall, D., Haarmeyer, D., Lindström, A., 2010. The Impact of the Global Financial Crisis on Financial Flows to the Water Sector in Sub-Saharan Africa. Stockholm, SIWI.

Kapdi, S.S., Vijay, V.K., Rajesh, S.K., Prasad, R., 2005. Biogas scrubbing, compression and storage: perspective and prospectus in Indian context. Renewable Energy 30(10), 1195-1202.

Karak, T., Bhattacharyyab, P., 2011. Human urine as a source of alternative natural fertilizer in agriculture: A flight of fancy or an achievable reality. Resources, Conservation and Recycling 55(4), 400-408.

Kaseva, M.E., Gupta, S.K., 1996. Recycling-an environmentally friendly and income generation activity towards sustainable solid waste management. Case study Dar es Salaam city. Resource conservation and Recycling 17, 299-309.

Kaseva, M.E., Mbulingwe, S.E., 2003. Appraisal of solid waste collection following private sector involvement in Dar es Salaam city, Tanzania. Habitat International 29(2), 353-366.

Katukiza, A.Y., Ronteltap, M., Niwagaba, C., Kansiime, F., Lens, P.N.L., 2010. Selection of sustainable sanitation technologies for urban slums — A case of Bwaise III in Kampala, Uganda, Science of the Total Environment 409(1), 52-62.

KCC (Kampala City Council), 2006. Solid waste management strategy report. Kampala City Council (KCC), Uganda.

Knerr, H., Engelhart, M., Hansen, J., Sagawe, G., 2008. "Separated grey- and black water treatment by the KOMPLETT water recycling system - A possibility to close the domestic water cycle". The Sanitation Challenge, 19th to 21st May; 2008, Wageningen, The Netherlands.

Koné, D., Strauss, M., 2004. Low-cost options for treating faecal sludges (FS) in developing countries — challenges and performance. In: Liénard A, Burnett H, editors. Proceedings of the 6th International Conference on Waste Stabilisation

Pond and 9th International Conference on Wetland Systems, Avignon (France), pp 213-219.

Kujawa-Roeleveld, K.K., Zeeman, G., 2006. Anaerobic treatment in decentralized and source-separation-based sanitation concepts. Reviews in Environmental Science and Bio/Technology 5(1), 115-139.

Kulabako, N.R., Nalubega, M., Thunvik, R., 2007. Study of the impact of land use and hydrogeological settings on the shallow groundwater quality in a peri-urban area of Kampala , Uganda . Science of the Total Environment 381(1-3), 180-199.

Kulabako, N.R., Nalubega, M., Wozei, E., Thunvik, R., 2010. Environmental health practices, constraints and possible interventions in peri-urban settlements in developing countries – a review of Kampala, Uganda. International Journal of Environmental Health Research 20(4), 231-257.

Kulabako, R., Kinobe, J., Mujunga, J., Olwenyi, S., Sleytr, K., 2009. Grey water use in peri-urban households in Kitgum, Uganda. ROSA Publications - Sustainable Sanitation Practice 1(1), 1-10.

Kvarnström, E., Bracken, P., Kärrman, E., Finnson, A., Saywell, D., 2004. People-centred approaches to water and environmental sanitation. 30th WEDC International Conference. WEDC Vientiane, Lao PDR.

Kvarnström, E., Emilsson, K., Stintzing, A.R., Johansson, M., Jönsson, H., af Petersens, E., Schönning, C., Christensen, J., Hellström, D., Qvarnström, L., Ridderstolpe, P., Drangert, J., 2006. One step towards sustainable sanitation. EcoSanRes Publications Series(1). www.ecosanres.org/pdf_files/Urine_Diversion_2006-1.pdf [Accessed on 21st October, 2010].

Langergraber, G., Muellegger, E., 2005. Ecological Sanitation-a way to solve global sanitation problems? Environment International 31, 433-444.

Larsen, T.A, Maurer, M., 2011. Source Separation and Decentralization. In: Wilderer P, editor. Treatise on Water Science, Oxford: Academic Press p. 203-229.

Leeming, R., Bate, N., Hewlett, R., Nicols, P.D., 1998. Discriminating faecal pollution: a case study of storm water entering port Philip Bay, Australia. Water Science and Technology 38(10), 15-22.

Lehmann, J., Gaunt, J., Rondon, M., 2006. Bio-char sequestration in terrestrial ecosystems-a review. Mitigation and Adaptation Strategies for Global Change 11(2), 395-419.

Lens, P.N., Vochten, P.N., Speleers, L., Verstraete, W.H., 1994. Direct treatment of domestic wastewater by percolation over peat, bark and wood chips. Water Research 28(1), 17-26.

Lettinga G., 2008. The dream of a clean environment for all (?) Proceedings of International IWA Conference (19-21 May 2008). Wageningen, the Netherlands.

Lettinga, G., Lens, P., Zeeman, G., 2001. Environmental protection technologies for sustainable development. Decentralized sanitation and reuse concepts, system and implementations, First ed. IWA publishing Alliance House, London, UK.

Li, F., Wichmann, K., Otterpohl, R., 2009. Review of technological approaches for grey water treatment and reuses. Science of the Total Environment 407(11), 3439-3449

Li, Z., Gulyas, H., Jahn, M., Gajurel, D., Otterpohl, R., 2003. Grey water treatment by Constructed wetlands in combination with TiO_2-based photocatalytic oxidation for suburban and rural areas without sewer system. Water Science and Technology 48(11-12), 101-106.

Lind, B.B., Bán, Z., Bydén, S., 2000. Nutrient recovery from human urine by struvite crystallization with ammonia adsorption on zeolite and wollastonite. Bioresource Technology 73, 169-174.

Lopez Zavala M.A., Funamizu, N., Takakuwa, T., 2004. Temperature effect on aerobic biodegradation of faeces using sawdust as a matrix. Water Research 38(9), 2405-2415.

Lüthi, C., McConville, J., Kvarnström, E., 2009. 'Community-based approaches for addressing the urban sanitation challenges', International Journal of Urban Sustainable Development 1(1), 49-63.

Maksimović, C., Tejada-Guibert, J.A., 2001. Frontiers in Urban Water Management. First ed. IWA-publishing, Alliance House, 12 Caxton street, London, SWIH 0QS, UK.

Mang, H.P,, Li, Z.O., 2010. Technology review "Biogas sanitation". Program Sustainable sanitation – ecosan of Deutsche Gesellschaft für Technische Zusammenarbeit (GTZ) GmbH.

Mara, D., Alabaster, G., 2008. A new paradigm for low-cost urban water supplies and sanitation in developing countries. Water Policy 10, 119-129.

Mara, D., Drangert, J.O., Anh, N.V., Tonderski, A., Gulyas, H., Tonderski, K., 2007. Selection of sustainable sanitation arrangements. Water Policy 9, 305-318

Mara, D., Lane, J., Scott, B., Trouba, D., 2010. Sanitation and Health. PLoS Medicine 7(11), 1-7.

Mara, D.D., 2003. Water, sanitation and hygiene for the health of developing nations. Public Health 117(6), 452-456.

Mara, D.D., 2008. Sanitation now: What is good practice and what is poor practice? Proceedings of International IWA Conference 19-21 May; 2008. Wageningen, the Netherlands.

Martens, W., Böhm, R., 2009. Overview of the ability of different treatment methods for liquid and solid manure to inactivate pathogens. Bioresource Technology 100(2), 5374-5378.

Mata-Alvarez, J., Macé, S., Llabrés, P., 2000. Anaerobic digestion of organic solid wastes. An overview of research achievements and perspectives. Bioresource Technology 74, 3-16.

Meinzinger, F., Oldenburg, M., 2008. Characteristics of source-separated household wastewater flows- a statistical assessment. Proceedings of International IWA Conference 19-21 May; 2008. Wageningen, the Netherlands.

Metcalf and Eddy, 2003. Wastewater Engineering: Treatment and Reuse. 4th ed. New York, NY 10020, USA: McGraw-Hill Companies, 1221 Avenue of Americas.

Moe, C.L., Rheingans, R.D., 2006. Global challenges in water, sanitation and health. Journal of Water and Health 04. Suppl, 41-57.

Morel. A., Diener, S., 2006. Greywater Management in Low and Middle-Income Countries. Review of different treatment systems for households or neighbourhoods.
http://www.eawag.ch/organisation/abteilungen/sandec/publikationen/publicati ons_ewm/downloads_ewm/Morel_Diener_Greywater_2006.pdf [Accessed on 11th September, 2010].

Mueller, C., 2007. Anaerobic Digestion of Biodegradable Solid Waste in Low- and Middle-Income Countries. Swiss Federal Institute of Aquatic Science (EAWAG), Department of Water and Sanitation in Developing Countries (SANDEC).

Mugo, K., 2006. Sustainable water and sanitation services for the urban poor in Nairobi. Paper presented at WEDC International Conference. 32nd WEDC, International Conference on sustainable development of water resources, water supply and environmental sanitation; Colombo, SriLanka.

Müller, C., 2007. Anaerobic Digestion of Biodegradable Solid Waste in Low- and Middle-IncomeCountries. Overview over existing technologies and relevant case studies. Eawag-SANDEC publications.

Mungai, G.G., 2008. Impacts of long-term greywater disposal on soil properties and reuse in urban agriculture in an informal settlement – A Case Study of Waruku, Nairobi. MSc Thesis, UNESCO-IHE Institute for Water Education, Delft, The Netherlands.

Nakagawa, N., Oe, H., Otaki, M., Ishizaki, K., 2006. Application of Microbial risk assessment on a residentially-operated Bio-toilet. Journal of Water and Health 4(4), 479-486.

Neves, L., Gonçalo, E., Oliveira, R., Alves, A.A., 2008. Influence of composition on the biomethanation potential of restaurant waste at mesophilic temperatures. Waste Management 28, 965-972.

Ngoc, U.N., Schnitzer, H., 2009. Sustainable solutions for solid waste management in South Asian countries. Waste Management 29, 1982-1995.

Niwagaba, C., Kulabako, R.N., Mugala, P., Jönsson, H., 2009. Comparing Microbial die-off in separately collected faeces with ash and sawdust additives. Waste Management 29(7), 2214-2219.

NWSC (National Water and Sewerage Corporation), 2008. Kampala Sanitation Program (KSP) - Feasibility study for sanitation master in Kampala, Uganda.

Nyenje, P.M., Foppen, J.W., Uhlenbrook, S., Kulabako, R., Muwanga, A., 2010. Eutrophication and nutrient release in urban areas of sub-Saharan Africa: a review. Science of the Total Environment 408(3), 447-455.

Ogawa, M., Okimori, Y., Fumio, Takahashi, F., 2006. Carbon Sequestration by Carbonization of Biomass and Forestation: Three Case Studies. Mitigation and Adaptation Strategies for Global Change 11(2), 421-436.

Ogwueleka, T.Ch., 2009. Municipal solid waste characteristics and management in Nigeria. Iran. J. Environ. Health. Sci. Eng. 6(3), 173-180.

Okot-Okumu, J., Nyenje, R., 2011. Municipal solid waste management under decentralisation in Uganda. Habitat International 35, 537-543.

Ondieki, T., Mbegera, M., 2009. *Impact assessment report on the PeePoo bag, Silanga village, Kibera, Nairobi-Kenya.* Bonn, Germany, Deutsche Gesellschaft fur Technische Zusammenarbeit (GTZ). www2.gtz.de/Dokumente/oe44/ecosan/en-peepoo-bags-assessment-Kibera-2009.pdf. [Accessed on 28th August, 2009].

Otterpohl, R., Albold, A., Oldenburg, M., 1999. Source control in urban sanitation and waste management: ten systems with reuse of resources. Water Science and Technology 39(5), 153-160.

Ottoson, J., Stenström, T.A., 2003. Faecal contamination of grey water and associated microbial risks. Water Research 37(3), 645-655.

Panuvatvanich, A., Koottatep, T., Kone, D., 2009. Influence of sand layer depth and percolate impounding regime on nitrogen transformation in vertical-flow constructed wetlands treating faecal sludge. Water Research 43, 2623-2630.

Paraknowitsch, J.P., Thomas, A., Antonietti, M., 2009. Carbon colloids prepared by hydrothermal carbonization as efficient fuel for indirect carbon fuel cells. Chem. Mater 21, 1170-1172.

Parkinson, J., Tayler, K., 2003. Decentralised wastewater management in peri-urban areas in low-income countries. Environ. Urbanisat. 15(1), 75-89.

Paterson, C., Mara, D., Cutis, T., 2007. Pro-poor sanitation technologies. Geoforum 38(5), 901-907.

Paterson, C., Mara, D.D., Curtis, T., 2007. Pro-poor sanitation technologies. Proceeding of Sanitation Challenge: New Sanitation Concepts and Models of Governance, 260-269, Wageningen, The Netherlands.

Pathak, B., 2009. World water week-opening plenary session Stockholm Water Prize Laureate.
http://www.sulabhinternational.org/downloads/pathak_opening_plenary_sessi on_sulabh_17aug09.pdf [2009, Oct.18].

Prüss, A., Kay, D., Fewtrell, L., Bartram, J., 2002. Estimating the burden for disease for water, sanitation, and hygiene at global level. Environ Health Perspect 110(5), 537-542.

Rajabapaiah, P., 1998. Energy from Bangalore garbage- A preliminary study. ASTRA technical report, CST. Bangalore.

Rauch, T., Drewes, J.E., 2005. Quantifying Biological Carbon Removal in Groundwater Recharge Systems. Journal of Environmental Engineering 131(6), 909-923.

Reijnders, L., 2009. Are forestation, bio-char and land filled biomass adequate offsets for the climate effects of burning fossil fuels? Energy Policy 37(8), 2839-2841.

Reith, J.H., Wijffels, R.H., Barten, H., 2003. Bio-methane & Bio-hydrogen: Status and perspectives of biological methane and hydrogen production. Dutch Biological Hydrogen Foundation, the Netherlands. ISBN: 90-9017165-7.

Ridderstolpe, P., 2007. Mulch Filter and Resorption for Onsite Grey Water Management- a report from the demo-facility built in Kimberly, South Africa. WRS 2007-02-12.

Rieck, C., von Münch, E., 2009. Technology review "Urine-diversion dehydration toilets". Sustainable sanitation - ecosan program of Deutsche Gesellschaft für Technische Zusammenarbeit (GTZ) GmbH.

Ronteltap, M., Maurer, M., Gujer, W., 2007. Struvite precipitation thermodynamics I source- separated urine. Water Research 41(5), 977-984.

Rutstein, S.O., 2000. Factors associated with trends in infant and child mortality in developing countries during the 1990s. Bulletin of the World Health Organization. 78(10), ISSN 0042-9686.

Schönning, C., Stenström, T.A., 2004. Guidelines for the Safe Use of Urine and Faeces in Ecological Sanitation Systems. EcoSanRes Publication Series. Report 2004-1. SEI; Stockholm, Sweden. http://www.ecosanres.org/pdf_files/ESR-factsheet-05.pdf [Accessed on 25th September, 2010].

Schouten, M.A.C., Mathenge, R.W., 2010. Communal sanitation alternatives for slums: A case study of Kibera, Kenya. Physics and Chemistry of the Earth 35(13-14), 815-822.

Sharholy, M., Ahmad, K., Vaishya, R.C., Gupta, R.D., 2007. Municipal solid waste characteristics and management in Allahabad, India. Waste Management 27, 490–496.

Sidhu, J.P.S., Toze, S.G., 2009. Human pathogens and their indicators in biosolids: A literature review. Environment international 35, 187-201.

Siegrist, H., Witt, M., Boyle, W.C., 1976. Characteristics of rural household wastewater. Journal of the Environmental Engineering Division 102(3), 533-548.

Sinnatamby, G.S., 1990. Low cost sanitation. In: Hardoy JE, Cairncross S, Satterthwaite D, editors. The Poor Die Young. Earthscan, London.

Siobhan, J., Rodda, N., Salukazana, L., 2006. Microbiological assessment of food crops irrigated with domestic greywater. Water SA 32(5), 700-704.

Steinbeiss, S., Gleixner, G., Antonietti, M., 2009. Effect of biochar amendment on soil carbon balance and soil microbial activity. Soil Biology & Biochemistry 41(6), 1301-1310.

Still, D.A., Salisbury, R.H., Foxon, K.M., Buckley, C.A., Bhagwan, J.N., 2010. The challenges of dealing with full VIP latrines. Proceedings WISA Biennial Conference & Exhibition, Durban ICC, South Africa, 18-22 April 2010. Water Research Commission Report No. TT 357/08, ISBN 978-1-77005-718-0.

Strauss, M., Drescher, S., Zurbrügg, C., Zurbrügg, A., 2003. Co-composting of Faecal Sludge and Municipal Organic Waste. SANDEC / EAWAG- IWMI; A Literature and State-of-Knowledge Review.

Styen, M., Jagals, P., Genthe, B., 2004. Assessment of microbial infection risks posed by ingestion of water during domestic water use and full-contact recreation in a mid-southern Africa region. Water Science and Technology 50(1), 301-308.

Sulabh International Social Service Organisation, 1999. Sanitation is the key to healthy cities – a profile of Sulabh international. Environment and Urbanisation 11(1), 221-230.

Surendran, S., Wheatley, A.D., 1998. Grey-water reclamation for non-potable re-use. Water and Environment Journal 12(6), 406–413.

Thye, Y.P., Templeton, M.R., Ali, M., 2011. A Critical Review of Technologies for Pit Latrine Emptying in Developing Countries, Critical Reviews in Environmental Science and Technology 41(20), 1793-1819.

Tilley, E., Lüthi, C., Morel, A., Zurbrügg, C., Schertenleib R., 2008 Compendium of Sanitation Systems and Technologies, EAWAG.

Tilley, E., Zurbrügg, C., Lüthi, C., 2010. A Flowstream Approach for Sustainable Sanitation Systems. In Social Perspectives on the Sanitation Challenge. Eds. Van Vliet B, Spaargaren G, Oosterveer P. Springer, p. 69-86.

Titirici, M.M., Thomas, A., Yu, S-H., Müller, J-O., Antonietti, M., 2007. A direct synthesis of mesoporous carbons with bicontinuous pore morphology from crude plant material by hydrothermal carbonization. Chem. Mater. 19, 4205-4212.

Udert, K.M., Larsen, T.A., Gujer, W., 2003. Estimating the precipitation potential in urine-collecting systems. Water Research 37, 2667–2677.

UN-Habitat, 2006. Rejuvenation of Community toilets. Policy paper 3 pp 1-14.

Victora, C.G., Smith, P.G., Vaughan, J.P., Nobre, L.C., Lombard, C., Teixeira, A.M.B., Fuchs, S.C., Moreira, L.B.B., Gigante, L.P., Barros, F.C., 1998. Water Supply, Sanitation and Housing in Relation to the Risk of Infant Mortality from Diarrhoea. International Journal of epidemiology 17(3), 651-654.

Vinnerås, B., 2007. Comparison of composting, storage and urea treatment for sanitising of faecal matter and manure. Bioresource Technology 98(17), 3317-3321.

Vinnerås, B., Hedenkvist, M., Nordin, A., Wilhelmson, A., 2009. Peepoo bag: self-sanitising single use biodegradable toilet. Water Science and Technology 59(9), 1743-9.

Vinnerås, B., Nordin, A., Niwagaba, C., Nyberg, K., 2008. Inactivation of bacteria and viruses in human urine depending on temperature and dilution rate. Water Research 42(15), 4067-4075.

Viswanath, P.S., Devi, S.N., And, K., 1992. Anaerobic digestion of fruit and vegetable processing wastes for biogas production. Bioresource Technology 40(1):43-48.

von Felde, K., Kunst, S., 1997. N- and COD-removal in vertical-flow systems. Water Science and Technology 35(5), 79-85.

von Münch, E., 2008. Overview of anaerobic treatment options for sustainable sanitation systems. International Symposium "Coupling Sustainable Sanitation and Ground Water Protection". Hannover, Germany.

von Münch, E., Mayumbelo, K.M.K., 2007. Methodology to compare costs of sanitation options for low-income peri-urban areas in Lusaka, Zambia. Water SA 33(5), 593-602.

Water Environment Foundation (WEF) and the American Society of Civil Engineers (ASCE), 1998. Design of Municipal Wastewater Treatment Plants (4th ed.) WEF Manual of Practice 8, ASCE Manual and Report on Engineering Practice No. 76. Volume 3. Virginia: WEF.

WaterAid India, 2006. Sanitation for all – still a long way to go; learning and approaches. Second South Asian Conference on Sanitation (Sept. 20-21), Islamabad, Pakistan.

Watson, G., 1995. Good sewers cheap? Agency–customer interactions inlow-cost urban sanitation in Brazil. World Bank Water and Sanitation Division, Washington DC.

WHO and UNICEF, 2010. Progress on drinking water and sanitation. Joint Monitoring Program Report (JMP). 1211 Geneva 27, Switzerland.

WHO and UNICEF, 2012. Progress on drinking water and sanitation. Joint Monitoring Program Report (JMP). 1211 Geneva 27, Switzerland.

Winsley, P., 2007. Biochar and bioenergy production for climate change mitigation. New Zealand Science Review 64(1), 5-10.

Zuma, B.M., Tandlich, R., Whittington-Jones, K.J., Burgess, J.E., 2009. Mulch tower treatment system; Part I: overall performance in grey water treatment. Desalination 242, 38-56.

Chapter 3: Selection of sustainable sanitation technologies for urban slums

This Chapter is based on:

Katukiza AY, Ronteltap M, Niwagaba C, Kansiime F, Lens PNL., 2010. Selection of sustainable sanitation technologies for urban slums - A case of Bwaise III in Kampala, Uganda, *Science of the Total Environment* 409(1), 52-62.

Abstract

Provision of sanitation solutions in the world's urban slums is extremely challenging due to lack of money, space, access and sense of ownership. This paper presents a technology selection method that was used for selection of appropriate sanitation solutions for urban slums. The method used in this paper takes into account sustainability criteria, including social acceptance, technological and physical applicability, economical and institutional aspects, and the need to protect and promote human health and the environment. The study was carried out in Bwaise III; a slum area in Kampala (Uganda). This was through administering of questionnaires and focus group discussions to obtain baseline data, developing a database to compare different sanitation options using technology selection criteria and then performing a multi-criteria analysis of the technology options. It was found that 15% of the population uses a public pit latrine; 75% uses a shared toilet; and 10% has private, non-shared sanitation facilities. Using the selection method, technologies such as Urine Diversion Dry Toilet (UDDT) and biogas latrines were identified to be potentially feasible sanitation solutions for Bwaise III. Sanitation challenges for further research are also presented.

3.1 Introduction

3.1.1 Sanitation in slum areas

The major contributors of the pollution load in urban slums into the environment are excreta, grey water and solid wastes (Eriksson et al., 2002; Howard et al., 2003; Kulabako et al., 2007; Paterson et al., 2007; Eriksson et al., 2009). Slums in developing countries lack basic sanitation services due to poor accessibility, lack of legal status and financial resources (Maksimović and Tejada-Guibert, 2001; Kulabako et al., 2007; Paterson et al., 2007) as well as lack of supportive infrastructure (von Münch and Mayumbelo, 2007). The main sanitation challenges for slums are the ways of enhancing demand for sanitation, the sustainability question and the institutional structures and arrangements for upscaling and replication by other practitioners (Franceys, 2008; Jenkins and Sugden, 2006).

One of the ways to deal with pollution streams in urban slums is through the provision of well functioning sanitation systems. Sanitation here refers to the management of human excreta, grey water, solid waste and storm water. The main polluting constituents are pathogens that endanger public health and nutrients that may cause eutrophication of surface waters and pollution of groundwater. Human excreta management is key to public health in urban slums since most of the pathogens are of faecal origin. They form a major cause of disease transmission due to the presence of pathogens in excreta and when mixed with wastewater, the pathogens flow downstream and spread in the environment especially during flooding (Feachem et al., 1983; Prüss et al., 2002; Niwagaba et al., 2009).

Human excreta are predominantly disposed in slum areas by use of unlined pit latrines which are usually elevated in areas with a high water table. Other excreta disposal facilities and options include traditional pit latrines, flying toilets (use of polythene bags for excreta disposal that are dumped into the surrounding environment), open defecation and to a small extent ventilated improved pit latrines (VIP) and pour flush toilets by the few high income earners (Kulabako et al., 2007; Muwuluke, 2007). These excreta disposal systems in use are considered unimproved because they are shared by many households (WHO and UNICEF, 2010). Moreover, they pollute the groundwater through direct and indirect discharge of pollution loads into the environment.

The current need to increase sanitation coverage to meet the MDG sanitation target has triggered provision of sanitation systems also in slum areas. However, little thought is given to the treatment of the collected waste. The high population density in slums and the typical flood prone locations tend to enhance the problem of overflowing pit latrines and washout of flying toilets. Excreta disposal facilities are a component of the sanitation system, which can be sustainable if it is affordable, developed to ensure pathogen removal and efficient for the recovery and reuse of nutrients contained in the excreta (Otterpohl et al., 1999; Mara, 2003; Kvarnström and af Petersens, 2004; Langergraber and Muellegger, 2005; Bracken et al., 2007). Sustainability of sanitation systems in urban slums may be achieved if technology selection methods that take into account the local situation are adopted.

3.1.2 Case study area: Bwaise III

This study focused on Bwaise III (32° 34'E, 0° 21'N), a slum area located in Kampala, Uganda (Figure 3.1). Bwaise III has six Local Council (LC) 1 zones (the lowest administrative unit at the Local Government Level) namely: Kawaala, Katoogo, St. Francis, Bugalani, Bokasa, and Kalimali. In total, 15,015 people are estimated to live in an area of 57 ha making the population density 265 persons per hectare (UBOS, 2002; Kulabako et al., 2007). The population growth rate of the study area is currently estimated by Kawempe Divison to be 9%. This is above the national average value of 3.4%. The area was initially a wetland connected to the existing Lubigi wetland to which it drains.

The slum area evolved as a result of illegal encroachment on the wetland. Consequently, there is no legal status of ownership of the area which makes provision of basic water and sanitation services difficult. The high water table and vulnerability of the area to flooding in rainy seasons has also made it difficult to construct pit latrines that are largely used in this area for excreta disposal. This indicates that there is a strong need to take the location into account in providing appropriate sanitation. Recently, an extensive study was carried out to investigate the groundwater situation in Bwaise III (Kulabako et al., 2007; Kulabako et al., 2008). It was concluded that a large part of the pollution was sanitation-related. This study therefore focuses on the sanitation situation in Bwaise III by analysing the current situation and developing a method to select sustainable sanitation options for excreta management.

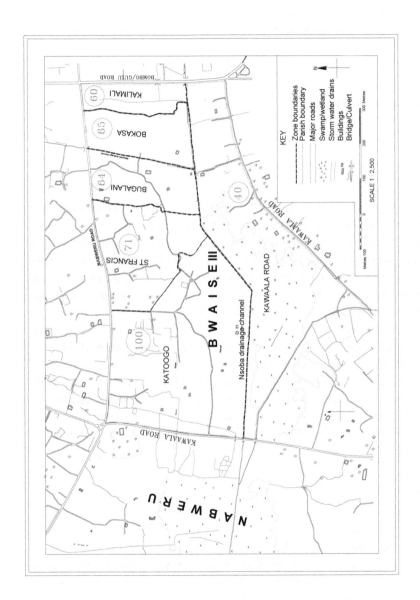

Figure 3.1: Map of Bwaise III parish showing the Local Council zones and the corresponding sample size indicated in circles

53

3.2   Materials and methods

This study was carried out in three phases. The first phase involved conducting a household survey to obtain baseline data. The second phase involved developing a database to enable comparison of different technology solutions based on established criteria to relate them to the baseline data. Finally, a multi-criteria analysis was carried out to incorporate the views of experts (technical and non-technical professionals) and stakeholders into the technology selection.

3.2.1   Sample size and selection

The sample size of 400 households was adopted for the study area of Bwaise III parish in Kawempe division. It was derived from the following expression developed for calculating a sample for proportions (Cochran, 1963; Kish, 1965):

$N = Z^2\left(\frac{Pq}{e^2}\right) = Z^2\left(\frac{P(1-P)}{e^2}\right)$, where N is the sample size, e is the desired level of precision ± 5% (Bryan, 1992; de Langen, 2007), Z is 1.96 at 95% confidence interval, P is the estimated level of an attribute present in the population which is 0.627 (toilet coverage of 62.7% from the Kawempe Division three year development plan 2006 – 2009) and q is 1-P. The expression gives a value of 361 but 400 households were used in this study to cater for none responses to the interviews and the 5% precision and sampling error. The number of households sampled in the parish zones of Kawaala, Katoogo, St. Francis, Bugalani, Bokasa, and Kalimali were 40, 100, 71, 64, 65 and 60, respectively. It was based on the number of households in a parish zone. The randomisation was based on the type of sanitation facility, the household income level (low, medium), the family head (male, female) and the need to cover different locations in an LC1 zone. The final decision on the actual number of households selected and their location in each zone took into consideration the actual situation on the ground through consultations with the Kawempe Division Administrators, Local Council Leaders and field observations. The study area boundaries were the six zones of Bwaise III parish (Figure 3.1).

3.2.2   Questionnaires

Structured, semi-structured and unstructured questionnaires were used to collect information from household heads and other key informants in Bwaise III. Unstructured questionnaires were based on the interaction with the informant to elicit information and used as part of observation field work (Punch 1998, Patton 1990). They were formulated to include sustainability issues and their indicators as applied to sanitation options in slum areas. The questionnaire was checked for validity by pre-testing it on 10% of the households that were used in the study after which improvements were made for final administering in the study area. Data collection was carried out either early in the morning or late in the evening being the time when most residents and household heads are at home (Muwuluke, 2007). Household heads were targeted rather than women as household income earners are the ones who make the final decisions. Of the families investigated, 30% are headed by women. In addition, there were polygamous families with only women available at home to respond to the questionnaires because their husbands were rarely present. Household interviews were also conducted by employing a random selection method and targeting household heads (Bryan, 1992) in order to obtain information on water consumption, excreta management systems including their ownership and solid waste management practices.

3.2.3 Field investigations

Data on the current sanitation practices and sanitation situation in Bwaise III was collected by interviewing key informants and through transect walks in the study area. Key informants are people and organisations that have been involved in the implementation of sanitation activities in the study area. They included officials from government institutions (Ministries and Universities), non-governmental organizations and private sector. Transect walks were also made in the six LC1 zones together with their respective LC1 leaders to assess the current sanitation practices and functionality of the existing sanitation facilities. These were systematic walks with key informants through the study area aimed at observing the current sanitation practices and carrying out informal and informative interviews using a checklist for completeness.

3.2.4 Technology selection process scheme

A flow chart (Figure 3.2) shows the method used to select the technically feasible and sustainable sanitation options. It includes two levels: technology selection and multi-criteria analysis.

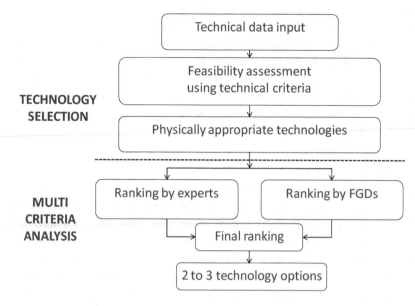

Figure 3.2: Flow chart for selection of sustainable sanitation options applied in this study (FGDs: focus group discussions, MCA: multi-criteria analysis)

3.2.4.1 Technology Selection

Technology selection was carried out using an Excel-based tool that was developed under this study. It comprises of the input data that are area specific, an assessment sheet where technology characteristics are subjected to the technical criteria and environment compliance as well as the output sheet with technically feasible options. The aim was to eliminate non-feasible options for the next level of assessment. Area specific input data for the following verifiable and policy variables were:

a) Data on the following known and verifiable parameters: present population density, accessibility, data on existing water supply situation, ground water table and excavation constraints.

b) Variables for parameters that can change according to strategic decisions or technologies e.g. persons served, service levels and coverage. These are targets set by the service providers like National Water and Sewerage Corporation (NWSC). Moreover, they also have to be in line with the Government policies such as the policy on water supply to the urban poor at lower tariff and the Poverty Eradication Action Plan (PEAP, 2004).

A number of technologies were considered in the technology selection, based on whether they are currently used in slums or have potential for use in slum settings. Conventional sewerage is costly to implement and operate and is not suitable for urban slum settings. It has been found to be an anti-poor technology by virtue of its higher cost and water requirements compared to simplified sewerage, which has been used in urban poor areas in South Asia, South America and South Africa (Paterson et al., 2007; Mara, 2003). Waterborne systems are feasible where the communities can afford the related tariffs, there is space for the simplified sewerage network including the wastewater treatment unit and the resource recovery in form of nutrients and biogas is not a priority. There are also prefabricated systems from polyethylene such as mobilets but their functionality is the same as that of pit latrines which is one of the technology options. Therefore, the technology options that were subjected to technical feasibility and further screening by multi-criteria analysis for applicability in slum areas included: urine diversion dry toilet (UDDT), biogas toilet/latrine, compost pit latrine, traditional pit latrine, lined ventilated improved pit latrine, pit latrine with urine diversion, *Fossa Alterna*, pour flush toilet connected to twin pits and simplified sewerage possibly connected to the main sewer of the nearby urban conventional system.

The identified technology options were subjected to technical criteria to determine their appropriateness in the study area. They included: water availability and consumption for water borne systems, excavation depth, accessibility to vacuum trucks and pickups and treatment requirements such as recovery of nutrients and energy in the form of biogas and environment protection against pollution based on National and World Health Organization effluent discharge standards. It was assumed that the maximum flood level is 0.5m above ground level, that the space between housing blocks is adequate for implementation of these systems and that there is market for the sanitised end products.

3.2.4.2 Multi-criteria analysis

The selected technologies were screened further by use of multi-criteria analysis (MCA) to take into account the perception of the stakeholders. The selected sanitation options for excreta disposal were presented to the stakeholders for ranking. This was done using focus group discussions (FGDs) taking into account gender, age and representation from the six zones of the study area. There were a total of six FGDs; three for both females and males each composed of representatives from two neighbouring LC 1 zones in Bwaise III. In addition, various experts (n = 20) participated in the ranking of the technically feasible sanitation options.

3.2.4.3 Ranking of technologies by FGDs

The pair-wise method was used for ranking of the sanitation technologies by the FGDs on a pair by pair basis. Using this structured method, two sanitation technologies were compared each at a time for the nine technology options (Table 3.1A). FGDs were held to establish perceptions and preferences from the communities about the technology options suitable for the study area (Table 3.1B). The technologies were presented to the beneficiary community (represented by FGDs) using IEC (Information, Education and Communication) materials (Figure 3.3) in the participatory discussion of the merits and demerits of these technologies with respect to sustainability indicators before the ranking activity.

In each comparison, the technologies were given scores on a scale of 1 to 5 (1 being the lowest and 5 the highest) based on their attributes and this varied for different FGDs (Table 3.3B). The FGD ranking for a particular technology was based on the FGD count referred to as the number of times it was chosen as the preferred option in the pair-wise matrix (Table 3.1A) using the sustainability criteria (Table 3.2). The FGD counts for a particular technology from the six FGDs were added to obtain its total FGD count used for the overall ranking (Table 3.1B). For technologies that have the same total FGD count, a comparison between the two in the pair-wise matrix was used to break the tie. In this study, a compost pit latrine and a biogas latrine ranked the same because they were each preferred by 3 FGDs in the matrix.

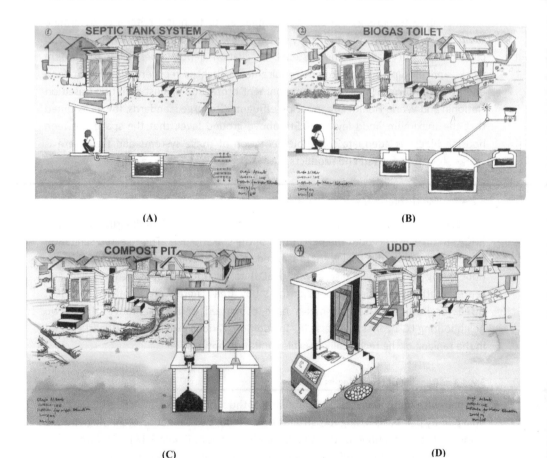

(A)   (B)

(C)   (D)

Figure 3.3:   Information, Education and Communication (IEC) materials used in the focus group discussions: (A) Septic tank system, (B) Biogas toilet, (C) Compost pit latrine and (D) Urine Diversion Dry Toilet (UDDT)

This method was more useful in prioritizing the needs of the beneficiary community than in a one-stage ranking. It offers an opportunity for special interest groups such as women, children, elderly and people with disabilities to influence the choice and design of the sanitation technologies. This is in agreement with Jenkins and Sugden (2006) that the beneficiaries should be involved at early stages of the design process and systems must be designed so that they meet their needs. Pair-wise ranking enabled the potential beneficiaries to make decisions in a consensus oriented manner.

Table 3.1A: **Matrix for ranking of technologies using the Pair-wise method. This table represents the outcome of a Focus Group Discussion (FGD) of females in Katoogo and Kawaala road Local Council zones**[*]

	(a)	(b)	(c)	(d)	(e)	(f)	(g)	(h)	(i)	FGD count	FGD ranking
(a)		b	c	d	e	f	a	h	i	1	8
(b)	b		c	d	e	b	b	h	i	3	5
(c)	c	b		d	e	f	c	c	i	3	6
(d)	d	d	d		e	d	d	d	i	6	4
(e)	e	e	e	e		e	e	e	i	7	2
(f)	f	f	f	d	f		f	f	i	6	3
(g)	g	b	c	d	e	f		g	i	2	7
(h)	a	b	c	d	e	f	g		i	0	9
(i)	i	i	i	i	i	i	i	i		8	1

[*](a) Septic tank system, (b) Biogas toilet/latrine, (c) Pour flush toilet with twin septic pits, (d) Urine Diversion Dry Toilet, (e) Compost pit latrine, (f) Pit latrine with Urine diversion, (g) Simplified sewerage system (h) Traditional pit latrine, (i) VIP latrine.

Table 3.1B: Ranking of technology options by Focus Group Discussions (FGDs)[*]

Technology	FGD counts for technology options						Total count	Overall ranking
(a)	1	2	1	0	1	1	6	8
(b)	5	4	6	6	4	3	28	4
(c)	4	3	5	4	3	3	22	5
(d)	7	7	6	7	6	6	39	1
(e)	2	5	4	5	5	7	28	4
(f)	4	5	6	5	7	6	33	3
(g)	3	2	0	2	0	2	9	6
(h)	0	0	2	2	4	0	8	7
(i)	7	5	4	5	5	8	34	2

[*](a) Septic tank system, (b) Biogas toilet/latrine, (c) Pour flush toilet with twin pits, (d) Urine Diversion Dry Toilet, (e) Compost pit latrine, (f) Pit latrine with Urine diversion, (g) Simplified sewerage system (h) Traditional pit latrine, (i) VIP latrine.

Table 3.2: Sustainability indicators and criteria

Sustainability indicator	Criteria
Socio-culture	Acceptance: *Proportion of users unhappy with the proposed technology option.*
	Perception/complexity: *Ability of beneficiaries to participate in operation and maintenance.*
	Use-ability: *How easy it is to use the proposed facility as viewed by the intended beneficiary community*
Technical	Local labour: *Capacity of local contractors to undertake the associated technical works.*
	Robustness: *Sensitivity to improper use, durability and sensitivity to harsh environment.*
	Materials: *Availability of local materials for facility construction.*
	Fit existing system: *Upgradeability to suit the local infrastructural and physical conditions.*
Health & Environment	Environmental pollution: *Risk of emission of pollutants to the environment such as nutrients and organic matter*
	Exposure to pathogens: *Risk of negative health impact associated with pathogens and contact with excreta during system management*
Economics	Capital cost: *Investment's requirement for the system*
	Land: *Space required for the system to be constructed*
	Operation & maintenance: *Resources (time, money, energy) for the system to serve its design life*
	Resource recovery: *Possibility of nutrient recovery from proposed technology for agricultural use.*
	Energy: *biogas recovery*
Institutional	Adoptability: *The ability of the beneficiary to use the technology*
	Management: *System for overseeing that the facility serves its intended purpose*
	Policy (WATSAN): *Strategic decisions by the government to increase sanitation coverage and service level to the urban poor*

3.2.4.4   Ranking of technologies by experts

Experts composed of technical and non-technical professionals ranked the technically feasible sanitation options. They included social scientists, engineers, public health specialists and institutional experts. Pre-determined sustainability criteria (Table 3.2) together with a matrix for ranking of technologies were presented to experts who determined the scores for the sustainability indicators (Table 3.3A). To estimate scores for the selected criteria for the alternative options, a scale of 1-5 was adopted (Al-Kloub et al., 1997; Mendoza et al., 1999; von Münch and Mels, 2008). Technical information about available excreta management technologies was established as reported in the literature (Mara and Guimarães, 1999; Paterson et al., 2007; Tilley et al., 2008; van der Steen, 2008). Sanitation was viewed as a system comprising of collection, storage, treatment and reuse/disposal. The selection of the potential sanitation option was based on technical feasibility and sustainability considerations used in previous studies (Kvarnström et al., 2004; Starkl and Brunner, 2004; Myšiak, 2006; von Münch and Mayumbelo, 2007) with involvement of the beneficiaries. Selected sustainability aspects that were considered and whose criteria were established included: socio-cultural, technical, health and environmental, economical and institutional (Table 3.2). These sustainability aspects were given scores based on their respective specific indicators depending on their degree of importance.

Table 3.3A: Scoring of sustainability indicators by experts (n=20)

Sustainability Indicator	Experts					
	Social Scientists/ Economists	Technical/ Engineers	Institutional specialists	Public Health specialists	Average scores	Weighted average scores (%)
Socio-culture	4.6	4.6	4.6	4.2	4.5	24.9
Technical	3.4	3.4	3.2	3.6	3.4	18.8
Health & Environme	3.8	3.6	4.4	4.8	4.2	23
Economics	4.2	4.2	2.6	3	3.5	19.4
Institutional	2.6	2.6	2.4	2.4	2.5	13.9
Total						100

Table 3.3B: Scoring of sustainability indicators by Focus Group Discussions (FGDs)

Sustainability function		Criteria	Average FGD scores for sustainability indicators based on scale of 1 (low) to 5 (high)					Total indicator score
Indicator			Biogas pit. lat./toilet	UDDT	Compost pit. Lat.	Lat. with urine diversion	Lined VIP	
Socio-culture		Acceptance	3	4	3	4	4	18
		Perception/complex	2	4	3	4	4	17
		Use-ability	3	4	3	4	5	19
								54
Technical		Local labour	3	5	4	4	5	21
		Robustness	2	4	2	2	2	12
		Materials	3	3	3	3	4	16
		Fit existing system	4	4	3	4	5	20
								69
Health & Environment		Environmental pollution	4	5	4	3	2	18
		Exposure to pathogens	4	4	3	2	2	15
		Land requirement	4	5	2	2	2	15
								48
Economics		Capital cost	3	4	4	4	3	18
		Operation & maintenance	4	4	3	3	3	17
		Nutrient reuse	2	5	3	2	1	13
		Energy recovery	5	1	1	1	1	9
								57
Institutional		Adoptability (local level)	3	4	3	4	4	18
		Policy (WATSAN)	3	3	2	3	4	15
								33

3.2.4.5   Final ranking of the sanitation technologies

The final ranking was achieved using the average FGD scores for the parameters defining sustainability indicators and the weighted scores of the sustainability indicators by the experts. The normalized score of a sustainability indicator was obtained as follows:

$$F = \left[\sum_{i=1}^{n} \left(\frac{a_i}{c} \right) \right] \times G$$ where F is the normalized score of a sustainability indicator, n is the number of parameters defining the criteria for a sustainability indicator, a is the average FGD score of a parameter for sustainability criteria, c is the total of the average FGD scores for criteria defining a sustainability indicator and G is the expert's weighted score for a sustainability indicator. The sum of the normalized scores (F) for all sustainability indicators was the total final score for a technology option and determined its final rank.

3.3   Results

3.3.1   Existing sanitation situation

3.3.1.1   Human excreta management

The excreta disposal facilities in Bwaise III range from simple/traditional pit latrines to flush systems with septic tanks. The majority of the residents (>50%) use elevated pit latrines for excreta disposal. Other excreta disposal technologies used include Ventilated Improved Pit (VIP) latrines and pour flush toilets (Figure 3.4A). Most of the sanitary facilities are shared and 75% of the slum dwellers use shared facilities. Community or public toilets are accessed by 15% of the people and 10% of the existing sanitary facilities are privately owned and not shared Figure 3.4B). Shared facilities are used by at least two households responsible for their maintenance while public or community toilets are used by the general public located at market places and public institutions. The findings showed that almost all respondents claimed to have access to a sanitation facility irrespective of the type of system.

During the field visits, the user load of the pit latrines was found higher than expected. Transect walks through the area clearly revealed that the user-load per latrine was high, with almost all latrine facilities shared by more than two households. The user to stance ratio was found to range from 1:30 to 1:70 which is higher than the recommended value of 1:20 by the Uganda Ministry of Health. This is in agreement with a previous study by PLAN (2001) that found it common for 10 households to share a stance. The high usage frequency requires also a high pit latrine emptying frequency. In a situation where landlords are not held liable, this remains challenging. Similar experiences are reported in urban slums of other developing countries (Chaggu, 2004; Mugo, 2006). Several contributing factors to ground water pollution were identified: unlined pits of excreta disposal facilities, leachate from decomposing solid waste dumps, uncollected grey water and storm water. It was also observed that most of the elevated pit latrines have a provision at the rear end for emptying the faecal sludge in either the adjacent storm water drain during rains or in excavated pits. This is a cheap practice by slum dwellers to make a hole through the feacal sludge chamber for pit latrine emptying. It is one of the potential exposure pathways through which the inhabitants get infected with pathogens present in the unsanitised feacal sludge discharged into the environment.

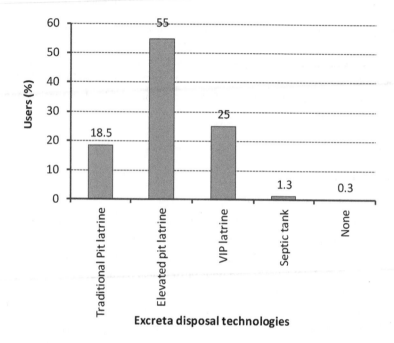

Figure 3.4A: Excreta disposal technologies in Bwaise III and the percentage of users making use of excreta disposal technologies

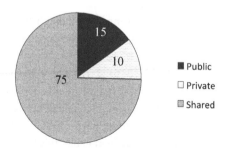

Figure 3.4B: Ownership of excreta disposal facilities in Bwaise III

3.3.1.2 Water supply situation

In Bwaise III, most residents have access to piped water from public standpipes. However, the cost per unit (*e.g.*, a 20 liter jerry can, or m^3) at the public standpipe is often more than 5 times higher than the tariff charged per m^3 by the National Water and Sewerage Corporation (NWSC) due to the extra charge by the service point caretakers in order to make profit. There are also a number of water vendors who sell a 20 litre jerry can of water at a price higher than that charged at standpipes (Oleja, 2009). They obtain water from the existing water sources to sell to consumers who lack a water connection and those who do not use unprotected springs as water sources due to public health concerns when there are dry zones with low pressure in the piped water supply network. Therefore, a number of slum dwellers who cannot afford this cost have resorted to obtaining water for potable use from nearby protected and unprotected springs and open shallow wells. The level of water supply service in slums is still low (Figure 3.5A) with very few house connections and an average per capita water consumption of 18 L/cap.d for the six local council zones (Figure 3.5B). This is lower than the minimum per capita water consumption of 20 L/cap.d set by the Uganda Ministry of Water and Environment (DWD, 2000) and the minimum global standards set by WHO.

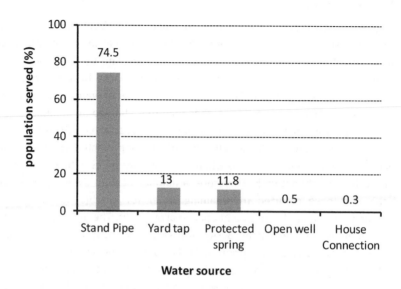

Figure 3.5A: Percentage of households using a water connection or water source

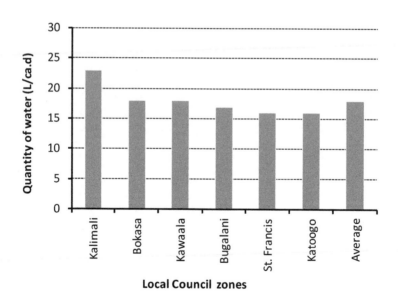

Figure 3.5B: Water consumption (L/capita.day) in the local council zones of Bwaise III

3.3.1.3   Solid waste management

There is poor solid waste management in Bwaise III which has resulted in illegal disposal of solid waste into storm water drains (Figure 3.6A) and next to the houses (Figure 3.6b), thus making the sanitation situation worse. The solid waste is largely organic, comprising food (bananas, potatoes, and pineapples) peelings, vegetable leaves and stems from households, small restaurants and food markets. Other fractions include plastics, glass and metal scrap.   The potential recycling options include compost manure, recycling of plastics, glass and metal scrap, and energy in the form of biogas (KCC, 2006). There are a number of stakeholders involved in the solid waste management that include the Kampala City Council (KCC) and private companies (Figure 3.7). There are also village health teams formed by the communities especially before the rainy season begins in response to common flooding problems. Households usually burn the solid waste during the dry season as a quickest way of disposing it. This is in spite of the negative effect to health by smoke and problematic noncombustible components of mixed waste such as fresh waste, glass and metallic parts though they account for a smaller fraction than the dry organic solid waste.

(A)   (B)   Elevated
                                                               pit latrine

Figure 3.6:   Photographs illustrating sanitation status in Bwaise III in Kampala (Uganda). (A) Blocked storm water drain and (B) Solid waste dump and an elevated pit latrine in the vicinity of a household in a slum area

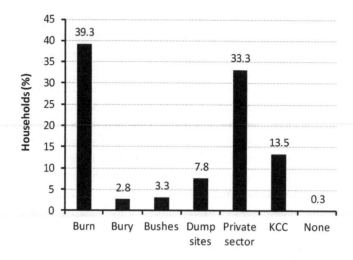

Solid waste disposal mechanism

Figure 3.7: **Typical Solid waste disposal mechanisms in Bwaise III and the Percentage of households applying a solid waste disposal mechanism**

3.3.2 Technology selection

Selection of appropriate technology for collection, storage, and treatment of waste streams is vital for sustainability of any sanitation system. In this study, a selection of the excreta disposal facilities was made with involvement of stakeholders to take into account their views (Figure 3.2). They included the residents of Bwaise III and professionals involved in the water and sanitation sector in Kampala. This was done after establishing the current sanitation situation in the study area in order to select sustainable sanitation systems. Physically appropriate technologies that passed the technical criteria were biogas latrine, Urine Diversion Dry Toilet (UDDT), compost pit latrine, pit latrine with urine diversion and lined ventilated improved (VIP) pit latrine. These were considered as physically appropriate for installation in the Bwaise III. The technologies that did not pass the technical criteria were the water borne systems including simplified sewerage pour-flush toilet and septic tank system due to low per capita water consumption and the unimproved technologies such as traditional pit latrine and *Fossa Alterna* that cause discharge of pollution load into the environment.

3.3.2.1 Focus Group Discussions

Focus Group Discussions (FGDs) were conducted to obtain the sanitation preferences of the residents and key informants. The residents were provided with information on the different sanitation options using Information, Education and Communication (IEC) materials as part of community participation in decision making. Table 3.1A shows a total count for each of the sanitation options by a FGD in each of the six zones of Bwaise III parish and this was done by the six FGDs. The total counts are results from pair wise ranking of the technologies by FGDs (Table 3.1A) based on a score scale of 1 to 5 with five being the highest. The most favoured option for use by the communities in the current situation was a UDDT followed by a lined VIP latrine (Table 3.1B).

3.3.2.2 Ranking of technology options by experts using sustainability indicators

Ranking of the selected options during multi-criteria analysis was carried out by professionals from academic institutions, government and private sector involved in the water and sanitation sector in Uganda (n = 20). This was based on various parameters considered as indicators for sustainability of the sanitation interventions in slum areas (Table 3.2). The results from the relevant professionals working in the water and sanitation sector (Figure 3.8) show that social scientists/economists and engineers ranked all aspects of sustainability to be almost equally important with the exception of health and environment. The average scores of the sustainability criteria by FGDs and the weighted scores of the sustainability indicators by experts were used to determine the final technology score and rank (Table 3.3B). A urine diversion dry toilet ranked the highest and a compost pit latrine ranked the lowest.

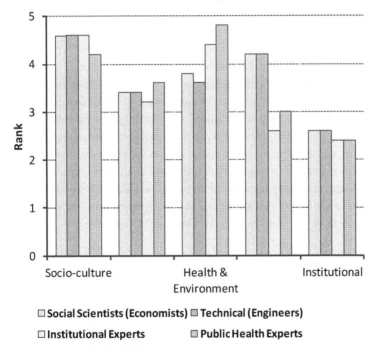

Figure 3.8:   Ranking of sustainability indicators by experts (n=20) during multi-criteria analysis

3.4   Discussion

3.4.1   Human excreta management

Pit latrines are the dominant type of excreta disposal in Bwaise III due to their low cost as a result of the use of available raw materials for construction and the low level of water services and lack of affordability for water borne systems. The pit latrines are elevated due to the high water table. Very few people use VIPs and flush toilets connected to the septic tanks due to the high investment costs of these systems and lack of willingness by landlords to invest in sanitation facilities. Use of polyethylene bags ("flying toilets") and open defecation with the former dumped in drainage channels or solid waste dumps has also aggravated the problem of pollution. Previous studies have reported that most of the spring water sources in the peri-urban areas of Kampala are contaminated with pathogens of faecal origin (Howard et al., 2003; Nsubuga et al., 2004; Kulabako et al., 2007), which can be attributed to the current poor sanitation practice.

There has been a lot of sanitation interventions (for excreta disposal) in Bwaise III by NGOs (PLAN, 2001) but most of them have failed due to lack of stakeholder participation at all stages of the project cycle. The lined VIPs that have been implemented were either based on systems implemented in areas with favourable soil formations such as non-collapsing soils with low water table or in other peri-urban areas that have space and legal status or ownership. Sanitation facilities in Bwaise III were constructed without considering sanitation as a system that comprises of collection, storage, treatment and safe disposal/reuse. The concept of a sanitation system was therefore lost and this caused failure to operate and maintain the systems, resulting in environmental pollution due to lack of treatment of increased volumes of waste generated. As a result, there are challenges to be overcome that include: design and implementation of flood proof sanitation systems suitable for local conditions in slums and developing an emptying system for many existing pit latrines (to be improved) in slums to cope with a high filling rate. In addition, further research is needed on comparative performance of improved mixed and separate excreta disposal systems to ascertain the attributes of the combinations with respect to pollution reduction in slums and recovery of resources in the form of nutrients and biogas.

The pit latrines in Bwaise III are shared by more than two households. The VIPs that were constructed by NGOs are shared by up to 7 households indicating a high user-load and consequently a high filling rate. Operation and maintenance of the shared toilets in Bwaise III is done by landlords, who reside in the affluent areas in Kampala and often outside the country and mainly aim at maximizing profit from rent while the few privately owned are maintained by the owners. This implies that current and future interventions in selecting technologies and subsequent provision sanitation services should involve Landlords, preferably through the Local Council Admnistration.

There is a problem of lack of willingness by the tenants to contribute to the maintenance of the shared facility. It is attributed to lack of ownership and the population dynamics: the population settlement versus occasional flooding and movement to look for better employment opportunities. The slum inhabitants whose income improves either become landlords in Bwaise III or in another place, or shift and rent in a better place (Kulabako et al., 2010). Consequently, the demographic dynamics in urban slums are a challenge for sanitation improvement.

3.4.2   Water supply

Water supply is as key as provision of a sanitation facility to ensure a safe water chain in Bwaise III. Water is used for domestic consumption including drinking, bathing, washing hands and for transport in case of water borne systems such as flush toilets connected to septic tanks. Generally in Kampala city in which Bwaise is located, there is infrastructure for water supply but the residents in peri-urban settlements have limited access to piped water. The supply is unreliable due to problems of dry zones in the network that is compounded by high costs (5 times the tariff of the National Water and Sewerage Corporation) charged by the caretakers of the standpipes and vendors. Other sources of water include springs and open well in the neighbourhood (Figure 3.5A). Rain water harvesting is also practiced during the rainy season despite lack of reliable data. The other issues include long walking distances to the water sources and lack of a corridor for water pipes and other basic services in unplanned informal settlements. These findings are similar to those obtained in previous studies for other surrounding peri-urban areas in Kawempe Division in Kampala (Kiyimba, 2006; Habitat, 2007) and are comparable to other similar settings in developing countries (Coates et al., 2004). Unfortunately, the shallow ground water sources have been found to be contaminated with nutrients (N, P), thermotolerant coliforms (TTCs) and faecal streptococci (FS) originating from multiple sources of contamination (Howard et al., 2003; Nsubuga et al; 2004; Kulabako et al., 2007). The low per capita water consumption also implies that use of water borne sanitation systems such as simplified sewerage may not be feasible sanitation options for urban slums even though a piped water infrastructure may be available.

3.4.3   Solid waste management

Lack of proper solid waste management is another major problem to the environment and to public health in slums. In developing countries, less than 50% of the solid waste generated (80% of which is organic) in urban areas is collected centrally by the municipalities or private sector, with limited recycling or recovery of recyclable materials (Kaseva and Mbulingwe, 2003; Chanakya et al., 2009). This low collection efficiency is also reflected in the case of Bwaise III. Despite the focus on sanitation improvement by some NGOs and CBOs, the low collection efficiency of waste in urban areas including slums implies that a substantial amount of solid waste remains uncollected. This most likely results in environmental pollution and negative public health effects. Poor management of organic solid waste can lead to methane emissions from solid waste dumps as well as leaching of nutrients and organic matter into the natural environment (Chanakya et al., 2009; Holm-Nielsen et al., 2009). Therefore, the main issues to address are collection of the increasing volume of solid waste generated, the heterogeneity of the solid waste characteristics and the treatment and disposal/reuse methods. For future research, a similar technology selection method can be applied to solid waste management using a participatory approach for technology options such as composting, anaerobic digestion and for recycling of plastics and metal scrap.

3.4.4   Ranking of the technologies using the sustainability criteria

Ranking of the technologies by both the beneficiary community and the professional experts determined the preferred sustainable sanitation options. Nine technologies (Table 1 B) were ranked by the FGDs instead of five technologies that passed the technical criteria (Table 3C). This was done to find out if the technically omitted technologies would be ranked low by the FGDs after they were presented with the merits and demerits of each technology using IEC materials (Figure 3.3). Water borne systems (simplified sewerage, septic tank system) and unimproved facilities (pit latrine) were ranked low in comparison to preferences such as a UDDT (Table 1B). The FGDs ranked the UDDT first followed by a lined VIP latrine. The latter was also favoured probably because it is one of the common improved sanitation systems in place compared to flush toilets connected to septic tanks used by only a small percentage of the population (Figure 3.4B). In addition, a traditional pit latrine which is one of the common types of excreta disposal facilities in Bwaise III ranked second last. This implies that the communities are aware of the environmental pollution caused by existing unimproved excreta disposal facilities but have no option due to lack of funds and complexity of the slum settings. Consequently, involvement of stakeholders enabled a better selection of the sanitation options suitable for the local situation. It is also important to note that under the four categories of social scientists, engineers, public health specialists and institutional experts, social aspects ranked highest with an average score of 4.5 compared to institutional aspects with an average score of 2.5 (Table 3A). This outlines the importance of non-technical considerations in choosing a sanitation system. A weighing was applied in order to integrate the results of the FGDs and the experts to reflect the opinions of the stake holders in the final ranking of the technologies after multi-criteria analysis.

The UDDT system ranked highest because of its attributes that include: construction and repair with locally available materials and small land requirements, no constant water requirement for use, prolonged service life since it can be emptied for reuse, suitablility for flood prone areas due to non mixing of waste streams and odour control that is achieved through proper usage. However, there are challenges in using this type of facility in slums such as acceptability by the users, market for the end products, source of ash for odour and smell reduction as well as pH elevation for pathogen inactivation, the high filling rate leading to high emptying frequencies for urine and faeces in slums. The compost pit latrine ranked low due to the requirement of secondary treatment of the leachate, design considerations including operation and maintenance as well as start up requirements. The results show that different specialists have different perceptions on what weighs most for a sanitation system to be sustainable. The different perceptions of experts influenced the results on the one hand (especially the economical and health and environment criteria), but increased the external validity of the sustainability concept on the other hand. A biogas latrine ranked second because it also requires less land and has benefits of biogas while a VIP latrine was ranked third because it is also one of the systems that can be constructed with locally available materials and does not use water.

The technology selection method that has been used in this study is useful in obtaining sustainable sanitation options for a particular urban slum based on the local conditions. It also takes into account social aspects that are usually left out during service provision in slums. All the steps in this method are needed in order to choose sustainable sanitation solutions for urban slums in any geographical location. The method does not predict the relative effects of the changes in sustainability aspects for long term planning, but allows making a choice by combining technological and social sustainability.

3.5 Conclusions

The existing sanitation systems in Bwaise III are unsustainable and are largely unimproved, which leads to ground water pollution and unhygienic conditions. In addition, most residents do not benefit from solid waste services provided by the local government or private sector due to limited access in the slum and low levels of affordability. Currently there are very few house water connections and an average water consumption of 18 L/cap.d which makes water-based systems such as simplified sewerage unattractive. Results from a multi-criteria analysis show that stakeholders' influence on sanitation options is key for sustainability, though ranking of appropriate sanitation options was also influenced by the users' and experts' perception of the technologies. The method of selecting technically appropriate sanitation options and thereafter letting the beneficiaries choose from the options through focus group discussions ensures a participatory approach. The identified sustainable sanitation options for Bwaise III are Urine Diversion Dry Toilet (UDDT) and biogas latrines.

References

Al-Kloub, B., Al-Shemmeri, T., Pearman, A., 1997. The role of weights in multi-criteria decision aid, and the ranking of water projects in Jordan. European Journal of Operational Research 99(2), 278-288.

Bracken, P., Wachtler, A., Panesar, A.R., Lange, J., 2007. The road not taken: how traditional excreta and greywater management may point the way to a sustainable future. Water Science & Technology, Water Supply 7(1), 219-227.

Bryan, F.J.M., 1992. The design and analysis of research studies. New York: Cambridge University Press.

Chaggu, E.J., 2004. Sustainable Environmental Protection Using Modified Pit-Latrines. PhD Thesis. Wageningen University, The Netherlands.

Chanakya, H.N., Sharma, I., Ramachandra, T.V., 2009. Micro-scale anaerobic digestion of point- source components of organic fraction of municipal solid waste. Waste Management 29, 1306-1312.

Coates, S., Sansom, K., Kayaga, S., Srivinas, C., Narender, A., Njiru, C., 2004. Serving all urbanconsumers-A marketing approach to water services in low and middle-income countries. Book 3: PREPP-Utility consultation with the urban poor. WEDC, Loughborough University, UK.

Cochran, W.G., 1963. Sampling Techniques, 2nd Ed., New York: John Wiley and Sons, Inc.

de Langen, M., 2007. Engineering Economics-Hypothesis Testing. UNESCO-IHE Institute for Water Education, Delft, The Netherlands.

DWD (Directorate of Water Development), 2000. Water supply Design manual, MWLE. Kampala, Uganda.

Eriksson, E., Andersen, H.R., Madsen, T.S., Ledin, A., 2009. Grey water pollution variability and loadings. Ecological Engineering 35, 661-669.

Eriksson, E., Aufarth, K., Henze, M., Ledin A., 2002. Characteristics of grey wastewater. Urban Water 4, 85-104.

Feachem, R.G., Bradley, D., Garelick, H., Mara, D., 1983. Sanitation and Disease: health espects of excreta and wastewater management. Chichester, UK: John Wiley and sons.

Franceys, R., 2008. 'Running Institutions or Walking the Talk', Stockholm Water Week. http://www.worldwaterweek.org/documents/WWW_PDF/2008/thursday/K16_17/Richard_Franceys_Sanitation.pdf. (Accesses on Oct. 10, 2010).

Holm-Nielsen, J.B., Al Seadi, T., Oleskowicz-Popiel, P., 2009. The future of anaerobic digestion and biogas utilisation. Bioresource Technology 100, 5478-5484.

Howard, G., Pedley, S., Barret, M., Nalubega, M., Johal, K., 2003. Risk factors contributing to microbiological contamination of shallow groundwater in Kampala, Uganda. Water Research 37, 3421–9.

Jenkins, M.W., Sugden, S., 2006. Rethinking sanitation. Lessons and innovation for sustainability and success in the New Millennium. UNDP - sanitation thematic paper.

Kampala City Council (KCC), 2006. City council of Kampala final solid waste management strategy report. Undertaken by Kagga & Partners Ltd in association with Kwezi v3 Engineers Consultants.

Kaseva, M.E., Gupta, S.K., 1996. Recycling-an environmentally friendly and income generation activity towards sustainable solid waste management. Case study Dar es Salaam city. Resource conservation and Recycling 17, 299–309.

Kish, L., 1965. Survey Sampling. New York: John Wiley and Sons, Inc.

Kiyimba, J., 2006. Identifying the levels of satisfaction, need and representation-Citizens action survey report in Kawempe Division for improvement of people's lives. Implemented by Community Integrated Development Initiatives (CIDI) in partnership with Water Aid, Kampala, Uganda.

Kulabako, N.R., Nalubega, M., Thunvik, R., 2007. Study of the impact of land use and hydrogeological settings on the shallow groundwater quality in a peri-urban area of Kampala , Uganda . Science of the Total Environment 381(1-3), 180-199.

Kulabako, N.R., Nalubega, M., Thunvik, R., 2008. Phosphorus transport in shallow groundwater in peri-urban Kampala, Uganda: results from field and laboratory measurements. Environmental Geology 53(7), 1535-1551.

Kulabako, N.R., Nalubega, M., Thunvik, R., 2010. Environmental health practices, constraints and possible interventionsin peri-urban settlements in developing countries – a review of Kampala, Uganda. International Journal of Environmental Health Research 20(4), 231–257.

Kvarnström, E., af Petersens E., 2004. Open Planning of Sanitation Systems Report No. 2004-3. Stockholm Environment Institute, Stockholm, Sweden. www.ecosanres.org/pdf_files/ESR_Publications_2004/ESR3web.pdf [Accessed on 17th July, 2009].

Kvarnström, E., Bracken, P., Kärrman, E., Finnson, A., Saywell, D., 2004. People-centred approaches to water and environmental sanitation. 30th WEDC International Conference. WEDC Vientiane, Lao PDR.

Langergraber, G., Muellegger, E., 2005. Ecological Sanitation--a way to solve global sanitation problems? Environment International 31(3), 433-444.

Maksimović, C., Tejada-Guibert, J.A., 2001. Frontiers in Urban Water Management. First ed. London: IWA-publishing.

Mara, D.D., Guimarães, A.S.P., 1999. Simplified sewerage: potential applicability in industrialized countries. Urban Water 1(3), 257-259.

Mara, D.D., 2003. Water, sanitation and hygiene for the health of developing nations. Public Health 117(6), 452-456.

Mendoza, G.A., Macoun, P., Prabhu, R., Sukadri, D., Purnomo, H., Hartanto, H., 1999. Guidelines for applying multi-criteria analysis to the assessment or criteria and indicators. Tool box series No. 9. Centre for International Forestry Research, Jakarta, Indonesia. http://www.cifor.cgiar.org; 1999 [Accessed on 17th February 2009].

Mugo, K., 2006. Sustainable water and sanitation services for the urban poor in Nairobi. Paper presented at WEDC International Conference. 32nd WEDC,

International Conference on sustainable development of water resources, water supply and environmental sanitation; Colombo, SriLanka.

Muwuluke, Z.J., 2007. Planning sustainable sanitation systems. A case of human excreta management in peri-urban areas of Kampala. PhD Thesis, Makerere University, Kampala, Uganda.

Myšiak, J., 2006. Consistency of the results of different MCA methods: A critical review. Environment and planning. C, Government & policy 24, 257-277.

Niwagaba, C., Kulabako, R.N., Mugala, P., Jönsson, H., 2009. Comparing Microbial die-off in separately collected faeces with ash and sawdust additives. Waste Management 29(7), 2214-2219.

Nsubuga, F.B., Kansiime, F., Okot-Okumu, J., 2004. Pollution of protected springs in relation to high and low density settlements in Kampala-Uganda. Journal of Physics and Chemistry of the Earth 29, 1153-1159.

Oleja, A., 2009. Identification of sustainable sanitation options for a slum area in Kampala, Uganda. MSc Thesis, UNESCO-IHE Institute for Water Education, Delft, The Netherlands.

Otterpohl, R., Albold, A., Oldenburg, M., 1999. Source control in urban sanitation and waste management: ten systems with reuse of resources. Water Science and Technology 39(5), 153-160.

Paterson, C., Mara, D., Curtis, T., 2007. Pro-poor sanitation technologies. Geoforum 38(5), 901-907.

Patton, M.Q., 1990. Quantitative evaluation and research methods. Newbury Park: Sage publications.

PEAP (Poverty Eradication Action Plan), 2004. Ministry of Finance, Planning and Economic Development (Uganda).

PLAN, 2001. Feasibility study project preparation Bwaise III – Kawempe Division. Plan International (PLAN) field report on community, environment, sanitation and solid waste management, drainage and flood aspects.

Prüss, A., Kay, D., Fewtrell, L., Bartram, J., 2002. Estimating the burden for disease for water, sanitation, and hygiene at global level. Environ Health Perspect 110(5), 537-542.

Punch, K.F., 2005 Introduction to Social Research: Quantitative and Qualitative Approaches. SAGE Publications, 2005.

Starkl, M., Brunner, N., 2004. Feasibility versus sustainability in urban water management. Journal of Environmental Management 71, 245-260.

Tilley, E., Luthi, C., Morel, A., Zurbrugg, C., Schertenleib, R., 2008. Compendium of Sanitation Systems and Technologies. Eawag-Sandec, Dubendorf, 158 pp.

UBOS (Uganda Bureau of Statistics), 2002. National Housing Census 2002, Uganda.

UN-Habitat, 2006. Rejuvenation of Community toilets. Policy paper 3 pp 1-14.

Van der Steen, N.P, 2008. Technology Selection for Sanitation and Wastewater Treatment. UNESCO-IHE Institute for Water Education, Delft, The Netherlands.

von Münch, E., Mayumbelo, K.M.K., 2007. Methodology to compare costs of sanitation options for low-income peri-urban areas in Lusaka, Zambia. Water SA 33(5), 593-602.

von Münch, E., Mels, A., 2008. Evaluating various sanitation system alternatives for urban areas by multi criteria analysis - case study of Accra. International IWA Conference, Wageningen, The Netherlands.

WHO and UNICEF, 2010. Progress on drinking water and sanitation. Joint Monitoring Program Report (JMP). 1211 Geneva 27, Switzerland.

Chapter 4: Genomic copy concentrations of selected waterborne viruses in a slum environment in Kampala, Uganda

This Chapter is based on the modified version of:

Katukiza, A.Y., Temanu, H., Chung, J.W., Foppen, J.W.A., Lens, P.N.L., 2013. Genomic copy concentrations of selected waterborne viruses in a slum environment in Kampala, Uganda. *Journal of Water and Health* 11(2), 358-369.

Abstract

The presence of waterborne viruses in a slum environment where sanitation is poor is a major concern. However, little is known of their occurrence and genomic copy concentration in the slum environment. The main objective of this study was to determine the genomic copy concentrations of human adenoviruses F and G, Rotavirus (RV), Hepatitis A virus (HAV), Hepatitis E virus (HEV) and human adenovirus species A,C,D,E, and F (HAdV-ACDEF) in Bwaise III, a typical slum in Kampala, Uganda. To detect waterborne viruses, 41 samples from surface water, grey water and ground water were collected every other day for a period of three weeks from 30 sampling locations. The virus particles were recovered by glass wool filtration with elution using beef extract. DNA and RNA viruses were detected by the real time quantitative polymerase chain reaction (qPCR) and the reverse transcriptase-qPCR (RT-qPCR), respectively. HAdV-F and G were detected in 70.7% of the samples with concentrations up to 2.65×10^1 genomic copies per mL (gc mL^{-1}). RV and HAV were detected in 60.9% and 17.1% of the samples, respectively. The maximum concentration of RV was 1.87×10^2 gc mL^{-1}. In addition, 78 % of the samples tested positive for the HAdV-ACDEF, but all samples tested negative for HEV. Surprisingly, HAdV-F and G were detected in spring water used for domestic water supply. Moreover, rotavirus was also detected in ground water beneath pit latrines. The results showed that surface water, ground water and grey water in the slum environment were strongly polluted with viral contaminants. These new data are essential for assessing the risk of infections caused by viruses, and for understanding the effects of environmental pollution in slums.

4.1 Introduction

The health risks caused by viruses present in water and waste water streams to which slum dwellers are exposed are a major concern. Waterborne disease outbreaks in slums of developing countries are caused by enteric viruses, bacteria and protozoa (Carr, 2001; Jaykus, 1997; Ashbolt, 2004; Montgomery and Elimelech, 2007; Bloomfield et al., 2009) and the situation is made worse by poor sanitation and hygiene practices (Howard et al., 2003; Jenkins and Sugden, 2006; Konteh, 2009; Mara et al., 2010). Outbreaks of typhoid fever, cholera, dysentery and diarrhoea are associated with the low level of sanitation and are the major causes of mortality of children under the age of five in slum areas (Legros et al., 2000; de Melo et al., 2008; Awasthi and Pande, 1998). Viruses, unlike bacteria, are very infective with one particle capable of causing an infection (Haas et al., 1993; Crabtree et al., 1997; Fong and Lipp, 2005). They are persistent in the environment and are resistant to conventional water and wastewater treatment technologies (Gantzer et al., 1997; Baggi et al., 2001; Fong et al., 2010). Enteric viruses can stay in the water matrix for months while remaining infective (Grabow, 1996; Jiang, 2006; Espinosa et al., 2008; Templeton et al., 2008).

The survival of virus particles in the environment is enhanced by the presence of fecal matter, colloidal clays, soils as well as biological and chemical floc particles (Templeton et al., 2008; Vasickova et al., 2010). Non-enveloped viruses are protected by a protein coat that interacts with the host cell surface and infectivity is dependent on whether the viral capsid is damaged or not (Templeton et al., 2008; Rodríguez et al., 2009). Infectivity of virus particles reduces with chlorination, exposure to sunlight, UV radiation and high temperature (Noble and Fuhrman, 1997; Carter, 2005; Espinosa et al., 2008; Rodríguez et al., 2009). Viruses are of public health concern due to their low infectious dose. The risk of infection from viruses is at least 10 fold greater than that for pathogenic bacteria, and 1 PFU (Plaque Forming Units) is capable of causing infection in 1% of healthy adults (Haas et al., 1993; Crabtree et al., 1997; Bosch, 1998; Fong and Lipp, 2005). In slums, the potential public health risk from pathogenic viruses may even be higher. The risk of infection from hepatitis A virus in surface water was found to be higher in communities with low socio-economic status (Venter et al., 2007).

Human adenovirus (HAdVs), rotavirus (RV), hepatitis A virus (HAV), noroviruses (NoV) and enteroviruses (EV) are some of the most prevalent viruses in wastewater and contaminated surface water (Schvoerer et al., 2000; van Zyl et al., 2006; Xagoraraki et al., 2007; Wyn-Jones et al., 2011). The sources of viruses include untreated or treated sewage or runoff that has been in contact with animal manure or human excreta. Enteric viruses are transmitted via the faecal-oral route (Brugha et al., 1999; van Zyl et al., 2006; Bloomfield et al., 2009). Gastroenteritis, conjunctivitis, diarrhoea and hepatitis are examples of diseases caused by viruses with morbidity and mortality occurring among children, the elderly persons and patients with low immunity

(Schvoerer et al., 2000; van Zyl et al., 2006; Xagoraraki et al., 2007). HAdV species F and G and RV are responsible for mortality in children and the elderly (Crabtree et al., 1997; Kocwa-Haluch and Zalewska, 2002; Rigotto et al., 2011). HAdV contain a linear, double-stranded DNA genome and are sometimes selected as a viral indicator of human faecal pollution because it is highly prevalent and stable (Muscillo et al., 2008; Pina et al., 1998; Rigotto et al., 2011; Wyn-Jones and Sellwood, 2001). RV are double-stranded RNA (dsRNA) viruses, with eleven genome segments covered within a three-layer capsid and are resistant to chloramines and ultraviolet light inactivation (Gerba et al., 1996). NoV, HAV and HEV are RNA viruses prevalent in Asia, China, Africa and South America, which pose a high risk of illness. In addition, the morbidity and mortality associated with HEV is significant for pregnant women (Wyn-Jones and Sellwood, 2001; van Cuyck et al., 2005; Caron and Kazanji, 2008; Dalton et al., 2008).

Previous studies in peri-urban areas in developing countries concluded that a large part of the pollution was sanitation related (Chaggu, 2002; Howard et al., 2003; Kulabako et al., 2007; Oswald et al., 2007; Genser et al., 2008; Mara et al., 2010). Boreholes, shallow wells, springs and surface water in slums and densely populated areas where the sanitary infrastructure is poor have been found to be contaminated with Escherichia coli, faecal coliforms and thermotolerant coliforms (Kimani-Murage and Ngindu, 2007; van Geen et al., 2011). Inadequate onsite and offsite sanitation systems are the principal causes of high pathogen concentrations in ground water and surface water of peri-urban areas (Karn and Harada, 2001; Paterson et al., 2007; van Geen et al., 2011). In developing countries, the major focus in solving sanitation related problems has been to reduce health risks rather than reduction of environmental impacts (Langergraber and Muellegger, 2005; Moe and Rheingans, 2006; Mara et al., 2010). The pollution load from slum areas, however, has the potential to result in both negative environmental and public health impacts as excreta disposal using onsite sanitation is inadequate.

Very limited studies have been carried out on the occurrence of enteric viruses in the environment in developing countries (Miagostovich et al., 2008; Rigotto et al., 2010; Verheyen et al., 2009), and even less in slum environments (Kiulia et al., 2010). In addition, public health and environmental specialists in developing countries generally lack information on the occurrence and concentrations of viruses in slum environments. This is likely hindering achievement of Millenium Development Goal 7; ensuring environmental sustainability (WHO and UNICEF, 2010). This study was conducted in Bwaise III, a typical slum in Kampala (Uganda) with very poor sanitation infrastructure. Our main objective was to determine the genomic copy concentrations (GC) of selected enteric viruses in various water and wastewater streams present in Bwaise III.

4.2 Materials and Methods

4.2.1 Characterisation of the slum environment

Temperature and rainfall data were obtained from the weather station at Makerere University within 1 km from the periphery of the study area. Discharge was monitored at the outlet of the Nsooba channel into which the Nakamilo channel flows (Figure 4.1) using a stream gauge. A diver (Schlumberger, Delft, The Netherlands) installed in the stream gauge was used to record semi-continuous water levels. A diver recording barometric pressure was used to correct the water level data for atmospheric pressure changes. The stream discharge Q (m^3s^{-1}) was then determined using the following rating curve:

$$Q = 0.0006H^2 - 0.0076H$$ ......(1)

where H is the stage or water level (cm). The rating curve was constructed from a series of discharge measurements which were determined in the field using the salt dilution method (Moore, 2003). The discharge and the corresponding water levels recorded by the diver were used to construct the rating curve.

4.2.2 Sample collection

A total of 41 samples from 30 sampling locations were collected between January 11 and February 3, 2011. Samples were collected from 25 locations in Bwaise III shown in Figure 4.1 and also from five surface water locations (B1, B2, B3, C2 and C3) in the neighbouring slums of Mulago and Kyebando. Samples from locations P1, P2, P3, P4, P5, P8, and P10 were collected 2-3 times to assess the temporal variability of virus presence and/or virus concentration.

Three types of samples were collected: 26 surface water samples, 11 grey water samples and four ground water samples (spring water and sample from beneath a pit latrine). The majority of the samples were collected from Bwaise III, which is a typical slum in sub-Saharan Africa. The criteria for study area selection included: informal settlement, poor sanitation infrastructure, low lying and with a high water table to be able to obtain ground water samples and the presence of main surface water drains originating from upstream of the slum, enabling the determination of effects of the slum on virus concentrations. Usually, six samples were collected from the area between 8-10 A.M. in new 10 L plastic containers that were reused after being washed three times with sodium hypochlorite (3.85% m/v), and rinsed with distilled water. The containers were also washed with water of which the sample was finally taken. The sampling locations were inlets, outlets and junctions of primary, secondary and tertiary drains, unprotected springs used as drinking water sources, and one groundwater observation piezometer (5 m deep and unplasticised PVC PN6 ND 50 mm, water table 1.5 m below ground level at the time of sampling). Primary and secondary drains convey a mixture of surface water and grey water while tertiary drains convey grey water from the bathrooms and kitchen verandas.

In addition to taking a sample in the field, temperature (°C), conductivity (EC) (μS cm^{-1}), dissolved oxygen (DO) (mg l^{-1}), and pH were measured immediately using a portable pH meter (pH 3310 SET 2, incl. a SenTix®41 probe,). After collection, the samples were stored at +4°C using ice blocks and transported to the Public Health Engineering Laboratory at Makerere University for processing within one hour.

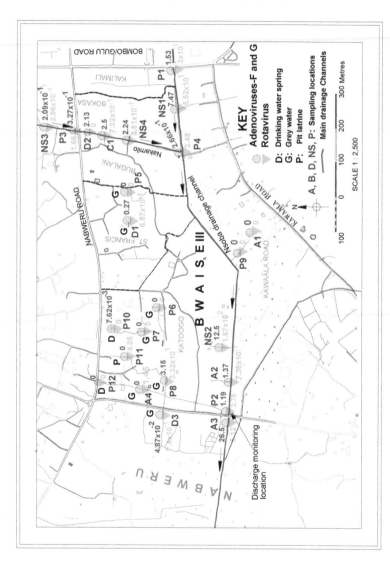

Figure 4.1: Genomic copy concentrations of human adenoviruses F and G (HAdV-F and G) and rotavirus (RV) (genomic copies mL^{-1}) in the Bwaise III slum environment

4.2.3   Virus concentration by glass wool filtration

To concentrate virus particles from the samples, we used a glass wool filtration protocol described by Wyn-Jones et al. (2011) which is a modification of the method of Vilaginès et al. (1993). Briefly, 10 L water samples were concentrated by adsorption at pH 3.5 to glass wool with a density of 0.25 gcm^{-3} and a total filter column height of 11.7 cm, and eluted with beef-extract/glycine buffer at pH 9.5 followed by organic flocculation at pH 4.5. The flocs were pelleted by centrifugation at 4200xg for 30 min and the pellet was resuspended in 10 mL of PBS. For our experiments, we used oiled white glass wool (Insulsafe 12 Isover Saint-Gobain, The Netherlands) and unplasticised PVC columns of 40 cm length and 30 mm internal diameter. Since this type of glass wool and the dimensions of the columns were similar, but not identical to the described protocol by Wyn-Jones et al. (2011), we first tested the virus recovery from our glass wool columns using known concentrations of the bacteriophages PRD1 and ϕX174. The stock solutions were kindly provided by the Microbiological Laboratory for Health Protection of the National Institute of Public Health and the Environment, where bacteriophage concentrations in the effluent samples were determined following standard protocols ISO 10705-1 (Anonymous, 1995) and ISO 10705-2 (Anonymous, 2000).

4.2.4   Nucleic acid extraction

Viral nucleic acids were extracted by a procedure described by Boom (Boom et al., 1999; Boom et al., 1990). The method is based on binding of nucleic acid to silica particles in the presence of a high molarity solution of guanidinium isothiocyanate (GSCN). Thereto, 500 μL of Boom buffer L7A (5.25 M GuSCN, 50 mM Tris-HCl [pH 6.4], 20 mM EDTA, 1.3% [wt/vol] Triton X-100, and 1 mg/mL alpha-casein) was added together with 10 μL micron sized silica particles suspension to 100 μL of sample concentrate in PBS (obtained from the previous step). After 30 minutes incubation at room temperature and centrifuging (at 13000 x g), the resulting silica pellet to which nucleic acids were adsorbed, was washed twice with Boom buffer L2 (5.25 M GuSCN, 50 mM Tris-HCl [pH 6.4]), twice with 70% ethanol and once with acetone. After centrifuging the silica pellet, disposal of the acetone, and drying of the pellet at 56 °C, 55 μL of TE buffer was added in order to desorb the nucleic acids from the silica particles. Finally, the vessel containing the silica pellet and TE buffer was vortexed 3 minutes, and incubated for 10 min at 56 °C. Then, the vessel was centrifuged for 2 min at 13,000 x g, and 50 μL of the supernatant was pipetted into a new Eppendorf. This fluid containing DNA and RNA was stored at -80 °C, and used for further experiments.

Table 4.1: Overview of primers, probes and thermocycler protocol used for the detection of targeted viruses

Virus	Oligo Name	Function	Sequences (5' ⟶ 3')	5'-Label	3'-Label	References
Adenovirus						
	Adeno-F	forward primer	CWTACATGCACATCKCSGG	-	-	(Hernroth et al., 2002)
	Adeno-R	reverse primer	CRCGGGCRAAYTGCACCAG	-	-	
	Adeno-probe	probe	CCGGGCTCAGGTACTCCGAGGCGTCCT	6FAM	BHQ1	
Adenovirus- E	Adeno-E	primers, probe	details for the primers and probe sequence with manufacturer(not provided)	6FAM	BHQ1	
Hepatitis -A						
	HAV240	forward primer	GGAGAGCCCTGGAAGAAAG	-	-	(Costafreda et al., 2006)
	HAV68	reverse primer	TCACCGCCGTTTGCCTAG	-	-	
	HAV150-Probe	probe	CCTGAACCTGCAGGAATTAA	6FAM	BHQ1	
Hepatitis-E						
	JVHEVF	forward primer	GGTGGTTTCTGGGGTGAC	-	-	(Bouwknegt et al., 2009)
	JVHEVR	reverse primer	AGGGGTTGGTTGGATGAA	-	-	
	JVHEV-Probe	probe	TGATTCTCAGCCCTTCGC	6FAM	BHQ1	
Rotavirus						
	Rota-NVP3-F	forward primer	ACCATCTACACTGACCCTC	-	-	(Pang et al., 2004)
	Rota-NVP3-R	reverse primer	GGTCACATAACGCCCC	-	-	
	Rota-Probe	probe	ATGAGCACAATAGTTAAAAGCTAACACTGTCAA	6FAM	BHQ1	

4.2.5 Quantitative PCR (qPCR) and Reverse transcription-qPCR (RT-qPCR)

The detection of DNA viruses was carried out using USB HotStart-IT Probe qPCR Master Mix (USB Cooperation, Cleveland, OH, USA), while the detection of RNA viruses was carried out using the Platinum Quantitative RT-PCR ThermoScript One-Step System (Invitrogen Life Technologies Cooperation, Grand Island, NY, USA). Primers and probe for HAdV-F and G were manufactured by Primerdesign Ltd (Southampton, UK), and were purchased as part of a kit ("Human Adenovirus Type F and G Standard Kit"). For HAdV-F and G, and HAdV-ACDEF, we used the thermocycler protocol as recommended by the USB HotStart-IT Probe qPCR Master Mix. Other primer and probe sequences including the thermocycler protocol were taken from literature (Table 4.1). All probes used consisted of a FAM fluorophore and a BHQ1 quencher and were manufactured by Biolegio (Nijmegen, The Netherlands). To each PCR well, 4 µL of sample was added to a total well volume of 25 µL. All qPCR and RT-qPCR analyses were carried out on a Mini-Opticon (Bio-Rad, Hercules, CA, USA).

4.2.6 Determining virus concentrations

Genomic copy concentrations of HAdV-F and G were determined with a standard curve (Figure 4.2), which was constructed in triplicate from a dilution series of a concentrated genomic copy number standard ($2x10^5$ copies μL^{-1}), that was included in the standard kit for quantification of human adenovirus type F and G (from Primerdesign). RV concentrations were determined with a standard curve (Figure 4.2), which was constructed in duplicate from a dilution series of concentrated genomic copy synthetic cDNA stock ($2.41x10^7$ copies μL^{-1}) with a length of 87 base pairs (position 963-1049; Pang et al., 2004) of the NSP3 Region; GenBank Accession Number X81436).

The results obtained by (RT)-qPCR were multiplied with a constant factor of $(1000x0.38)^{-1}$ to account for the up-concentration of the virus particles due to the glass wool protocol (from 10 L to 10 mL), and to account for the glass wool recovery of virus particles in the samples, which was based on the average recovery of the bacteriophages PRD1 and φX174 of 38% (see also 'Recovery of the bacteriophages' in the Results Section). The results were then multiplied by a factor of 0.5 to cater for the final recovered volume of RNA of 50 µL from the initial 100 µL of sample concentrate and by a dilution factor of either 10 or 100 for 16 samples that showed inhibition. The results of the target viruses HAdV-ACDEF, HAV, and HEV were reported in a qualitative way in terms of present or absent. Noroviruses were not investigated in this study, because RV is the leading cause of gastroenteritis and diarrhoea in children in developing countries (Clark and McKendrick, 2004; Ramani and Kang, 2009), who are among the vulnerable groups in slums.

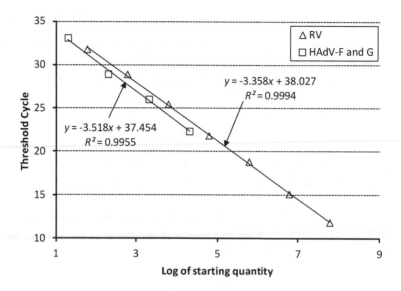

Figure 4.2: Standard curves of human adenovirus F and G (HAdV-F and G) and rotavirus (RV)

4.2.7 Inhibition and false negative tests

To check for inhibition of the PCR reaction, prior to detection of viral genomic copies, 4 µL of sample together with 4 µL of a known concentration of an artificially manufactured 80 nucleotides long single stranded piece of DNA of known composition (marker DNA3; Foppen et al., 2011) was added to a total volume of 25 µL, and then amplified according to the protocol described in Foppen et al. (2011).

4.2.8 Quality Control

Separate equipment, pipettes, filter tips, vials and reagent tubes were used for each stage of the process to avoid contamination of samples and reagents. Viral concentrates were kept frozen during transportation from Uganda to the laboratories of UNESCO-IHE (Delft, The Netherlands), where they were immediately stored at -80°C until further use. All qPCR and RT-qPCR assays included blank controls containing the same reaction mixture except for the nucleic acid template in addition to inhibition and false negative tests.

4.3 Results

4.3.1 Characterisation of Bwaise III

Bwaise III slum is located in Kawempe division (32° 34'E, 0° 21'N), which is one of the five city administrative units of Kampala (Uganda). It has an area of 57 ha with an altitude varying from 1166 to 1170 m above sea level. It is a reclaimed wetland, densely populated and experiences flooding during the rainy seasons (Katukiza et al., 2010). Rainfall data for the catchment showed a dry season during the sampling period from January 11 to February 3, 2011. There were a few minor rainfall events during this period and the daily ambient temperature varied from 23 °C to 28 °C. The minimum and maximum discharge at the outlet of the main drainage channel (Nsooba) recorded for the period from December 29, 2010 to February 8, 2011 were 0.15 and 1.54 m^3s^{-1}, respectively (Figure 4.3).

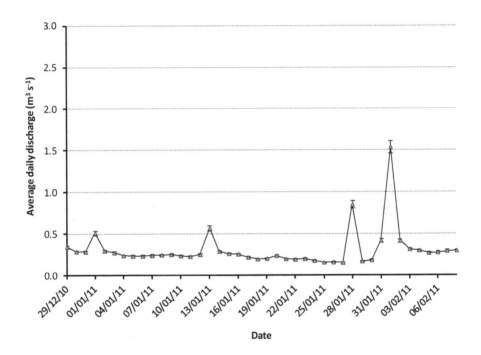

Figure 4.3: Average daily discharge (m^3s^{-1}) at the outlet of the Nsooba channel, the main drainage channel of Bwaise III slum area

4.3.2 Recovery of the bacteriophages

In our column setup, the recovery of bacteriophage PRD1 ranged from 12.2% to 26.3% and from 17.6% to 35.5%, when flushed with 10 L of tap water spiked with $2.0x10^2$ pfu mL^{-1} and $2.8x10^4$ pfu mL^{-1} phages, respectively. The recovery of φX174 ranged from 36% to 46.3% and 40% to 57.8% for tap water spiked with concentrations of 9 pfu mL^{-1} and $3x10^3$ pfu mL^{-1}, respectively. From this, we concluded that our column setup and the protocol used worked satisfactory. Two different concentrations were used to assess whether the glasswool protocol we applied yielded similar recovery rates for different concentrations, and also because we did not know the genomic copy concentrations we could expect in the various grey water, surface water and groundwater samples.

4.3.3 Sensitivity and efficiency of the HAdV-F and G and RV assays

For the HAdV-F and G assay, the coefficient of determination, R^2, was 0.995 (see Figure 4.2) and the efficiency of the qPCR reaction, determined as E = $[10^{-1/slope}]$ -1, was 0.92. For the RV assay, the coefficient of determination (R^2) was 0.999, while the efficiency of the RT-qPCR reaction was 1.23, which was high.

4.3.4 PCR inhibition tests

Out of 41 samples, 16 samples from 14 sampling locations showed inhibition of the PCR. This was 46% (12/26) of all surface water samples, 27% (3/11) of all grey water samples, and 25% (1/4) of all ground water samples. Out of four ground water samples, three spring water samples did not show PCR inhibition effects but one sample from beneath the pit latrine. To eliminate inhibition, 10 μL of the concentrated samples from the glass wool elution were diluted 10, 100, and 1000 times, and, after nucleic acid extraction, were again checked for inhibition. The 10 to 100 times dilution was in all cases adequate to eliminate inhibition of the PCR.

4.3.5 Prevalence of viruses in Bwaise III slum

Physicochemical parameters for the 4 groups of samples taken from surface water, spring water, grey water, and groundwater in the area differed from each other, while the standard deviations were relatively low (Table 4.2), indicating that the groups of waters we had sampled were relatively distinct and homogeneous from a physicochemical point of view. HAdV-ACDEF was detected in 78% of all samples (31/41; Table 4.3). All surface water samples contained HAdV-ACDEF, and so did 45% of the grey water samples. In addition, HAdV-ACDEF was found in one protected spring water sample, used for drinking water purposes. HAdV-F and G were detected in 70.7% of all samples (29/41): 96% of all surface water samples, 27% of grey water samples, and one spring sample that also contained HAdV-ACDEF (Table 4.3). RV was detected in 61% of all samples (25/41). Most of the surface water samples were positive for rotavirus, and so were 36% of the grey water samples. In addition, rotavirus was found in an observation piezometer sampling groundwater, almost beneath a pit latrine. HAV was found on few occasions (7/41), while HEV was not found at all. Finally, 17.1% (7/41) of the samples tested negative for all viruses. These were 2 spring water samples and 5 grey water samples. The distribution of viruses in the tested surface water, grey water and ground water samples is shown in 4.4.

Table 4.2: **Physico-chemical properties of tested samples**

Parameter	Samples and corresponding property values			
	Surface water (n = 26)	Spring water (n = 3)	Grey water (n = 11)	Ground water beneath a pit latrine (n = 1)
pH	7.1 (±0.2)	5.3 (±0.3)	7.5 (±0.4)	7.60
Temperature ($^{\circ}$C)	22.9(±1.4)	24.5(±1.5)	23.5 (±1.2)	25.2
EC (µS/cm)	575.4(±110)	613.5(±151)	1755.1 (±1443.4)	2480
DO (mg/L)	0.98 (±0.44)	4.17 (±0.54)	0.47(±0.34)	ND

Table 4.3: Occurrence of viruses in all tested surface water, ground water and grey water samples[*]

No.	Sampling location ID	Sampling location	Surface water					Grey water					Ground water				
			HAdV-F and G	AdV-ACDEF	RV	HAV	HEV	HAdV-F and G	AdV-ACDEF	RV	HAV	HEV	HAdV-F and G	AdV-ACDEF	RV	HAV	HEV
1	A1	Secondary drain															
2	A2	Nsooba	1.37	Prst	7.36×10^{-1}	Abst	Abst	0	Abst	0	Abst	Abst					
3	A3	Nsooba	2.65×10^{1}	Prst	5.12	Abst	Abst										
4	A4	Tertiary drain						0	Prst	1.45	Abst	Abst					
5	A5	Nakamilo	2.64×10^{-1}	Prst	1.04	Abst	Abst										
6	B1	Kyebando slum	1.09	Prst	0	Abst	Abst										
7	B2	Kyebando slum	0	Prst	0	Abst	Abst										
8	B3	Kyebando slum	0.69	Prst	0	Abst	Abst										
9	C1	Nakamilo	2.50	Prst	7.02×10^{-1}	Abst	Abst										
10	C2	Kyebando slum	4.41		6.63	Abst	Abst										
11	C3	Mulago slum	4.80		5.87×10^{-1}	Prst	Abst										
12	D1	Tertiary drain		Prst		Abst	Abst	2.7×10^{-1}	Prst	6.87×10^{-1}	Abst	Abst					
13	D2	Nakamilo	2.13	Prst	0	Abst	Abst										
14	D3	Tertiary drain						2.13	Prst	7.4×10^{-1}	Prst	Abst					
15	NS1	Nsooba	7.47	Prst	5.52×10^{1}	Abst	Abst										
16	NS2	Nsooba	1.25×10^{1}	Prst	1.87×10^{2}	Abst	Abst										
17	NS3	Nakamilo	4.18×10^{1}	Prst	2.96×10^{-1}	Abst	Abst										
18	NS4	Nakamilo	2.24	Prst	3.51×10^{-1}	Abst	Abst										
19	P1	Nsooba	1.53 (±1.1)	Prst	2.98×10^{1} (±3.66×10^{1})	Abst	Abst										
20	P2	Nsooba	1.19 (±0.2)	Prst	2.78 (±1.07)	Prst	Abst										
21	P3	Nakamilo	3.27×10^{-1} (±4.8×10^{-2})		1.66 (±5.63×10^{-1})	Prst	Abst										
22	P4	Nsooba	5.32×10^{-1} (±4.0×10^{-2})	Prst	2.48 (±9.61×10^{-2})	Prst	Abst										
23	P5	Tertiary drain						0	Abst	0	Abst	Abst					
24	P6	Tertiary drain						0	Abst	0	Abst	Abst					
25	P7	Tertiary drain						0	Abst	0	Abst	Abst					
26	P8	Tertiary drain						3.15 (±4.5)	Prst	3.32×10^{1} (±4.7×10^{1})	Prst	Abst					
27	P9	Secondary drain						0	Abst	0	Abst	Abst					
28	P10[†]	Spring water											7.62×10^{-3} (±1×10^{-2})	Prst	0	Abst	Abst
29	P11[††]	Piezometer											0	Abst	4.05	Abst	Abst
30	P12[†]	Spring water											0	Abst	0	Abst	Abst

[†]Ground water samples from spring water, [††]Sample from beneath the pit latrine

[*]Genomic copies ml^{-1} for HAdV-F and G, and RV; Prst (present)/Abst (Absent) for AdV-ACDEF, HAV and HEV. In brackets is the standard deviation

Table 4.4: Distribution of viruses in all tested surface water, groundwater and grey water samples

Virus type	Number (%) Viruses detected			
	Surface water (n = 26)	Spring water (n = 3)[*]	Grey water (n = 11)[**]	Ground water beneath a pit latrine (n = 1)
Adenovirus ACDEF	26 (100%)	1 (33%)	5 (45%)	0
Adenoviruses 40 and 41	25 (96.2%)	1 (33%)	3 (27.2%)	0
Rotavirus	20 (76.9%)	0	4 (36.4%)	1(100%)
Hepatitis A virus	4 (15.4%)	0	3 (27.2%)	0
Hepatitis E virus	0	0	0	0

[*]Two samples from the spring which acts as a drinking water source tested negative for all viruses

[**] Five grey water samples from tertiary drains tested negative for all viruses

4.3.6 Temporal virus concentration variations

In surface water samples, the temporal variation of the genomic copy concentration of HAdV-F and G and RV was limited (Table 4.5). In addition, except for P5, HAdV-ACDEF was always present. HAV was only occasionally present or absent. For the spring water and grey water samples, all viruses were occasionally present or absent.

Table 4.5: Temporal variation of virus presence and virus concentration for specific locations sampled two or three times (HEV was absent in all samples)

Virus type	Concentration of HAdV-F and G, and RV [†]							Sampling date
	Surface water				Spring water	Grey water		
	P1(n=3)	P2 (n=3)	P3 (n=3)	P4 (n=3)	P10*(n=2)	P5 (n=2)	P8 (n=2)	
Human adenoviruses- F and G	1.2	1.1	0.4	0.3	0	0	6.3	12 January 2011
	0.7	1.1	0.3	0.3	7.62×10^{-3}	0	0	19 January 2011
	2.7	1.4	0.3	0.3				27 January 2011
Rotavirus	0	3.84	2.02	2.55	0	0	6.64×10^{1}	12 January 2011
	1.88×10^{1}	1.71	1.95	2.41	0	0	0	19 January 2011
	7.06×10^{1}	2.78	1.01					27 January 2011
Human Adenovirus ACDEF	Prst	Prst	Prst	Prst	Abst	Abst	Prst	12 January 2011
	Prst	Prst	Prst	Prst	Abst	Prst	Abst	19 January 2011
	Prst	Prst	Prst	Prst				27 January 2011
Hepatitis A virus	Abst	Abst	Abst	Abst	Abst	Abst	Abst	12 January 2011
	Abst	Prst	Abst	Abst	Abst	Abst	Prst	19 January 2011
	Abst	Abst	Prst	Prst				27 January 2011

[*] P1, P2, P3, P4, P5, P8 and P10 refer to the sampling locations.

[†] Genomic copies ml^{-1} for adenoviruses- F and G, and rotavirus; Prst (present)/Abst (Absent) for human adenovirus ACDEF

4.3.7 Genomic copy concentrations (GC) of HAdV-F and G, and RV

We determined the genomic copy concentrations of HAdV-F and G and RV in surface water, spring water and grey water (Figure 4.1 and Table 4.3). The lowest concentration of HAdV-F and G for samples that tested positive was 7.62×10^{-3} gc mL^{-1} in a spring water sample, while the maximum was 2.65×10^{1} gc mL^{-1} in surface water at the outlet of the slum (Figure 4.1). In the surrounding slums, the highest concentration was 4.8 gc mL^{-1} indicating that HAdV-F and G was practically omnipresent. The genomic copy concentration of HAdV-F and G increased downstream of the two main storm water drains in the area. In the Nakamilo drain, the genomic copy concentration of HAdV-F and G increased from 2.09×10^{-1} gc mL^{-1} immediately North of Bwaise III to 2.24 gc mL^{-1} while in the Nsooba drain, it increased from 0.153×10^{1} gc mL^{-1} to 2.65×10^{1} gc mL^{-1}. In contrast, the genomic copy concentration of RV fluctuated along the drains without a clear pattern. The highest concentration of RV was 1.87×10^{2} gc mL^{-1} in surface water, while lowest concentration of RV was 2.96×10^{-1} gc mL^{-1} in surface water. When comparing the 3 groups of waters, virus concentrations were the highest in surface water followed by grey water, and then groundwater, either from springs or from the observation piezometer we sampled (Table 4.3).

4.4 Discussion

4.4.1 Virus recovery and PCR accuracy

The recovery of the bacteriophages PRD1 and φX174 between 12.2 and 57.8% by glass wool adsorption in this study was comparable to the 8 to 28% recovery of adenovirus serotype 41 (Lambertini et al., 2008), 34.2 to 78.2% recovery of adenovirus spiked in fresh water concentrated by glass wool followed by elution with beef extract (Wyn-Jones et al., 2011), and 40% recovery of enteric viruses from wastewater, ground water and surface water (Wolfaardt, 1995; Grabow, 1996; van Heerden et al., 2005; van Zyl et al., 2006; Venter et al., 2007). The recovery of φX174 with an average diameter of 27 nm and an iso-electric point between 4.6 and 7.8 was much higher than that of PRD1 with an average diameter of 62 nm and iso-electric point of 3 to 4. Apparently, bacteriophage φX174 attached more to glass wool than PRD1, as a result of the larger electrostatic repulsion of PRD1, compared to the less negatively charged φX174 (Vilaginès et al., 1993; Wyn-Jones et al., 2011).

RT-qPCR/qPCR was used because it was shown to be an accurate method with high sensitivity to detect viral RNA/DNA in a variety of environmental samples, despite the presence of inhibitors in samples and adsorption of viruses to particulate material (Jiang et al., 2005; Espinosa et al., 2008; Rodríguez et al., 2009; Wyn-Jones et al., 2011). Nucleic acid amplification inhibition effects exhibited by some samples during PCR were probably due to the presence of humic acids and other organic compounds and high bacterial concentrations. Most of the samples had no PCR inhibition probably because they were removed during the elution with PBS after glass wool filtration and nucleic acid adsorption on the silica colloids. In addition, the PCR inhibitors may not have adsorbed to glass wool during filtration (Van Heerden et al., 2005).

4.4.2 Waterborne viral contamination in the Bwaise III slum

The concentration of HAdV-F and G virus particles generally increased downstream of the two main drains (Nsooba and Nakamilo; Figure 4.1) in Bwaise III slum. There was an increase from 1.53 gc mL^{-1} at the inlet of the Nsooba drain to 26.5 gc mL^{-1} at the outlet, and an increase from 2.09x10^{-1} gc mL^{-1} upstream to 2.24 gc mL^{-1} downstream of the Nakamilo drain. These values were lower than the concentration of HAdV-F and G reported in polluted surface water ranging from 48 to 4.24x10^{4} gc mL^{-1} (Xagoraraki et al., 2007; Haramoto et al., 2010) and higher than the values obtained in other polluted environments (e.g. 1.03x10^{-5} to 3.23x10^{-3} gc mL^{-1}; Chapron et al., 2000; Jiang et al., 2005; van Heerden et al., 2005; Mena and Gerba, 2009). In addition, the concentration of HAdV-F and G in drinking water of 7.62x10^{-3} gc mL^{-1} from the spring (sampling point P10; Figure 4.1) obtained under this study was higher than 6.15x10^{-6}

to 7×10^{-5} gc mL^{-1} reported in non-chlorinated municipal drinking water (Lambertini et al., 2008), but comparable to 1.4×10^{-4} to 3×10^4 gc mL^{-1} of HAdVs in untreated surface water for domestic consumption (Xagoraraki et al., 2007; Rigotto et al., 2010).

We attributed the increase in the concentration of HAdV-F and G downstream of the drains to the high spatial density of inhabitants using unlined and elevated pit latrines discharging untreated waste water in the environment, especially along the Nakamilo drain. A dilution effect by the Nsooba channel is then responsible for the decrease in concentration HAdV-F and G virus particles after the junction of two drains in comparison to that in the tributary Nakamilo drain. We observed that this pattern was not exhibited by RV, which can be due to the differences in stability of RV compared to HAdV. HAdV remain more infectious and stable in the environment than RV due to their double-stranded DNA (Enriquez et al., 1995; Mena and Gerba, 2009; Rigotto et al., 2011; Wyn-Jones et al., 2011). This is also evident from reported decay rate coefficients of around 0.025 day^{-1} for HAdV-F and G compared to 0.36 day^{-1} for RV (Pedley et al., 2006). These values show that HAdV-F and G is more stable than RV although the survival rate also depends on intrinsic (chemical and biological) as well as extrinsic (temperature, light) factors (Espinosa et al., 2008; Rodríguez et al., 2009). Another reason for the observed RV patterns could be that the primary reaction step converting rotavirus RNA into DNA with reverse transcriptase enzyme could have affected the accuracy among the various samples tested for RV. The low concentration of RV and HAdV-F and G at the same location may be attributed to higher bacterial contamination and reduced stability of virus particles by sunlight compared to other locations (Espinosa et al., 2008; Rodríguez et al., 2009). The high concentration for both HAdV-F and G, and RV in surface water at the slum outlet (sampling location A3) may be attributed to the resultant pollution load into the main drain from upstream anthropogenic activities.

The high genomic copy concentrations of enteric HAdV-F and G, and RV suggest the presence of infective viruses although PCR does not distinguish between infectious and non-infectious virus particles. In addition, virus particles remain stable and persistent in the environment while free nucleic acid is unstable in the environment (Enriquez et al., 1995; Bosch, 1998; Meleg et al., 2006; Espinosa et al., 2008). The presence of virus particles in the environment is therefore a public health concern even though the correlation between infectivity and persistence of viruses is still a challenge (Hamza et al., 2009). Virus infectivity is important for quantifying the public health risks in the slum.

HAdV-ACDEF was detected in all surface water samples and 78% of all samples, which was high compared to the presence of other viruses. We think this was because they were more persistent in this faecally contaminated environment. HAV was detected in surface water and grey water, but not in ground water because it is associated with infections in the communities and is less stable in the environment compared to adenoviruses (Biziagos et al., 1988; Venter et al., 2007; Wyn-Jones et al., 2011). HEV is associated with sporadic infections and epidemics in areas with poor sanitation and weak public-health infrastructure (van Cuyck et al., 2005; Dalton et al., 2008). HEV was not detected in any sample probably because there were no related disease outbreaks in the study area at the time of this investigation and hence no HEV particles were discharged in the environment.

4.4.3 Possible interventions needed based on the findings

The variation of viral genomic copy concentrations in the major drains with continuous flow was attributed to different anthropogenic activities that affected the quantity and quality of surface water. In addition, grey water probably accounted for a large volume of the discharge in drains during the dry season. Flooding events during the rainy seasons together with spillage and uncontrolled discharge of pit latrine contents in the adjacent storm water drains may increase the exposure of slum dwellers to viruses and other pathogens. Exposure to viruses in slums can be through drinking water from contaminated springs, use of sand deposited on drains' beds for construction, children playing in or near drains and involuntary ingestion of water especially during flooding.

The viral contamination in Bwaise III slum requires interventions to minimise public health effects and also to reduce environmental pollution. The polluted existing springs that act as drinking water sources should be closed. We also think there is a need for creating public awareness with regard to the dangers of domestic potable use of water from contaminated springs and the risk of infection from contaminated surface waters. The sanitation infrastructure in the slum needs to be improved to prevent discharge of human excreta in the environment although this will likely be hindered by the lack of legal status for slums and their recognition by the Local Government. Other interventions could target reducing flooding effects that increase the risk of infection caused by overflowing drains and spillage of pit latrine contents.

4.5 Conclusions

This study found that 85.4% (35/41) of the samples tested positive for at least one of the investigated viruses, which indicated that the slum environment was polluted. Human Adenoviruses F species (serotypes 40 and 41) and G species (serotype 52), and Rotavirus were dominantly detected in the samples throughout the area investigated. The concentration range of HAdV-F and G was 7.62×10^{-3} gc mL^{-1} to 2.65×10^{1} gc mL^{-1} while the concentration range of rotavirus was 2.96×10^{-1} gc mL^{-1} to 1.87×10^{2} gc mL^{-1} for samples that tested positive. The presence of HAdV-F and G in the drinking water source constitutes a potential public health hazard that requires urgent attention by the relevant authorities. Detection of RV in a ground water sample from beneath a pit latrine was an indication of diffuse contamination of ground water with viral pathogens. The potential public health risks from these viruses are higher in slums because of high population density and environmental pollution from unlined and elevated pit latrines, which are mostly used for excreta disposal. Further studies should be conducted on viral infectivity, viral loadings and their periodic variations in a slum environment.

References

Anonymous, 1995. ISO 10705-1: Water quality—Detection and enumeration of bacteriophages-part 1: Enumeration of F-specific RNA bacteriophages. Geneva, Switzerland: International Organisation for Standardisation.

Anonymous, 2000. ISO 10705-2: Water quality. Detection and enumeration of bacteriophages - part 2: Enumeration of somatic coliphages. Geneva, Switzerland: International Organisation for Standardisation.

Ashbolt, N.J., 2004. Microbial contamination of drinking water and disease outcomes in developing regions. Toxicology 198, 229-238

Awasthi, S., Pande, V.K., 1998. Cause-specific Mortality in Under Fives in the Urban Slums of Lucknow, North India. Journal of Tropical Pediatrics 44, 358-361.

Baggi, F., Demarta, A., Peduzzi, R., 2001. Persistence of viral pathogens and bacteriophages during sewage treatment: lack of correlation with indicator bacteria. Research in Microbiology 152, 743–751.

Biziagos, E., Passagot, J., Crance, J.M., Deloince, R., 1988. Long-term survival of hepatitis A virus and poliovirus type 1 in mineral water. Applied And Environmental Microbiology 54, 2705-2710.

Bloomfield, S.F., Exner, M., Fara G.M., 2009. The global burden of hygiene-related diseases in relation to the home and community. An IFH expert review; published on http:// www.ifh-homehygiene.org.

Boom, R., Sol, C., Beld, M., Weel, J., Goudsmit, J., Wertheim-van Dillen, P., 1999. Improved silica-guanidiniumthiocyanate DNA isolation procedure based on selective binding of bovine alpha-casein to silica particles. Journal of clinical microbiology 37(3), 615-619

Boom, R., Sol, C.J., Salimans, M.M., Jansen, C.L., Wertheim-van Dillen, P.M., van der Noordaa, J., 1990. Rapid and simple method for purification of nucleic acids. Journal of clinical microbiology 28, 495-503.

Bosch, A., 1998. Human enteric viruses in the water environment: a mini review. Int. Microbiol 1(3), 191–196.

Bouwknegt, M., Rutjes, S.A., Reusken, C.B.E.M., Stockhofe-Zurwieden, N., Frankena, K., de Jong, M., de Roda Husman, A.M., van der Poel, W.H.M., 2009. The course of hepatitis E virus infection in pigs after contact-infection and intravenous inoculation. BMC veterinary research 5, 7.

Brugha, R., Vipond, I.B., Evans, M.R., 1999. A community outbreak of food-borne small round-structured virus gastroenteritis caused by a contamination of water supply. Epidemiol. Infect 122, 145–154.

Caron, M., Kazanji M., 2008. Hepatitis E virus is highly prevalent among pregnant women in Gabon, central Africa, with different patterns between rural and urban areas. Virology Journal 5, 158.

Carr, R., 2001. *Water Quality, Guidelines, Standards and Health*. Assessment of risk and risk management for water related diseases. Eds. Fewtrell L., Bartram J. ISBN 1 90022280, IWA Publishing.

Carter, M.J., 2005. Enterically infecting viruses: Pathogenicity, transmission and significance for food and waterborne infection. J. Appl. Microbiol 98, 1354-1380.

Chaggu, E.J., Mashauri, A., Van Buren, J., Sanders, W., Lettinga, J., 2002. PROFILE: Excreta Disposal in Dar es Salaam. Environment Management 30, 609-620.

Chapron, C.D., Ballester,N.A., Fontaine, J.H., Frades, C.N. & Margolin, A.B., 2000. Detection of Astroviruses, Enteroviruses, and Adenovirus Types 40 and 41 in Surface Waters Collected and Evaluated by the Information Collection Rule and an Integrated Cell Culture-Nested PCR Procedure. Applied And Environmental Microbiology 66, 2520–2525.

Clark, B., McKendrick, M., 2004 A review of viral gastroenteritis. Curr Opin Infect Dis. 17, 461-469.

Costafreda, M.I., Bosch, A., Pinto, R.M., 2006. Development, evaluation, and standardization of a real-time TaqMan reverse transcription-PCR assay for quantification of hepatitis A virus in clinical and shellfish samples. Applied and Environmental Microbiology 72, 3846-3855

Crabtree, K.D., Gerba, C.P., Rose, J.B., Haas, C.N., 1997. Waterborne Adenovirus: A risk assessment. Water Science and Technology 35, 1-6.

Dalton, H.R., Bendall, R., Ijaz, S., Banks, M., 2008. Hepatitis E: an emerging infection in developed countries. Lancet Infect Dis 8, 698–709

de Melo, M.C.N., Taddei, J.A.A.C., Diniz-Santos, D.R., Vieira, C., Carneiro, N.B., Melo, R.F. & Silva, L.R., 2008 Incidence of Diarrhea in Children Living in Urban Slums in Salvador, Brazil. The Brazilian Journal of Infectious Diseases 12, 89-93.

Enriquez, E.C., Hurst, C.H., Gerba, C.P., 1995. Survival of enteric adeno viruses 40 and 41 in tap, sea and wastewater. Water Research 29, 2548-2553.

Espinosa, A.C., Mazari-Hiriart, M., Espinosa, R., Maruri-Avidal, L., Méndez, E., Arias, C.F., 2008. Infectivity and genome persistence of rotavirus and astrovirus in groundwater and surface water. Water Research 42, 2618-2628.

Fong, T.T, Phanikumar, M.S., Xagoraraki, I., Joan B. Rose, J.B., 2010. Quantitative Detection of Human Adenoviruses in Wastewater and Combined Sewer Overflows Influencing a Michigan River. Applied and Environmental Microbiology 76, 715–723.

Fong, T.T., Lipp, E. K., 2005. Enteric viruses of humans and animals in aquatic environments: health risks, detection and potential water quality assessment tools. Microbiology and Molecular Biology Reviews 69, 357-371.

Foppen, J.W., Orup, C., Adell, R., Poulalion, V., Uhlenbrook, S., 2011 Using multiple artificial DNA tracers in hydrology. Hydrol. Process. 25, 3101–3106.

Gantzer, C., Senouci, S., Maul, A., Levi, Y., Schwartzbrod, L., 1997. Enterovirus genomes in wastewater: concentration on glass wool and glass powder and detection by RT-PCR. Journal of Virological Methods 65, 265–271.

Genser, B., Strina, A., dos, Santos, L.A., Teles, C.A., Prado, M.S., Cairncross, S., Barreto, M.L., 2008 Impact of a city-wide sanitation intervention in a large urban centre on social, environmental and behavioural determinants of childhood diarrhoea: analysis of two cohort studies. International Journal of Epidemiology 37, 831-840.

Gerba, C.P., Rose, J.B., Haas, C.N., Crabtree, K.D., 1996. Waterborne rotavirus: a risk assessment. Water Research 30, 2929-2940.

Grabow, W. 1996. Waterborne diseases: Update on water quality assessment and control. Water S A 22, 193-202.

Haas, C.N., Rose J.B., Gerba, C.P., Regli, R., 1993. Risk assessment of viruses in drinking water. Risk Anal 13, 545–552.

Hamza, I.A., Jurzik, L., Stang, A., Sure, K., Überla, K., Wilhelm, M., 2009. Detection of human viruses in rivers of a densly-populated area in Germany using a virus adsorption elution method optimized for PCR analyses. Water Research 43, 2657-2668.

Haramoto, E., Kitajima, M., Katayama, H., Ohgaki, S., 2010. Real-time PCR detection of adenoviruses, polyomaviruses, and torque teno viruses in river water in Japan. Water Research 44, 1747–1752.

Hernroth, B.E., Conden-Hansson, A.-C., Rehnstam-Holm, A.-S., Girones, R., Allard, A.K., 2002. Environmental factors influencing human viral pathogens and their potential indicator organisms in the blue mussel, Mytilus edulis: The first Scandinavian report. Applied and Environmental Microbiology 68, 4523-4533.

Howard, G., Pedley, S., Barret, M., Nalubega, M., Johal, K., 2003. Risk factors contributing to microbiological contamination of shallow groundwater in Kampala, Uganda. Water Research 37, 3421–9.

Jaykus, L. 1997. Epidemiology and Detection as Options for Control of Viral and Parasitic Food borne Disease. Emerging Infectious Diseases 3, 529-539.

Jenkins, M.W., Sugden S., 2006. *Rethinking sanitation: lessons and innovation for sustainability and success in the new millennium.* UNDP Human Development Report.

Jiang, S., Dezfulian, H., Chu, W., 2005. Real-time quantitative PCR for enteric adenovirus serotype 40 in environmental waters. Can. J. Microbiol 51, 393-398.

Jiang, S.C., 2006. Human Adenoviruses in Water: Occurrence and Health Implications: A *Critical Review.* Environ. Sci. Technol 40, 7132 7140.

Karn, S.K. & Harada H., 2001. Surface Water Pollution in Three Urban Territoriesof Nepal, India, and Bangladesh. Environmental Management 28, 483–496.

Katukiza, A.Y., Ronteltap, M., Oleja, A., Niwagaba, C.B., Kansiime, F., Lens, P.N.L., 2010. Selection of sustainable sanitation technologies for urban slums - A case of Bwaise III in Kampala, Uganda. Science of The Total Environment 409(1), 52-62.

Kimani-Murage, E.W., Ngindu, A.M., 2007. Quality of Water the Slum Dwellers Use: The Case of a Kenyan Slum. Journal of Urban Health: Bulletin of the New York Academy of Medicine 84, 829-838.

Kiulia, N.M., Netshikweta, R., Page, N.A., van Zyl, W.B., Kiraithe, M.M., Nyachieo, A., Mwenda, J.M., Taylor, M.B., 2010. The detection of enteric viruses in selected urban and rural river water and sewage in Kenya, with special reference to rotaviruses. Journal of Applied Microbiology 109, 818-828.

Kocwa-Haluch, R., Zalewska, B., 2002. Presence of Rotavirus hominis in Sewage and Water. Polish Journal of Environmental Studies 11, 751-755.

Konteh, F.H., 2009. Urban sanitation and health in the developing world: reminiscing the nineteenth century industrial nations. Health Place 15, 69-78.

Kulabako, N.R., Nalubega, M., Thunvik, R., 2007. Study of the impact of land use and hydrogeological settings on the shallow groundwater quality in a peri-urban area of Kampala , Uganda. Science of the Total Environment 381, 180-199.

Lambertini, E., Spencer, S.K., Bertz P.D., Loge F.J., Kieke, B.A., Borchardt, M.A., 2008. Concentration of Enteroviruses, Adenoviruses, and Noroviruses from Drinking Water by Use of Glass Wool Filters. Applied and Environmental Microbiology 74, 2990-2996.

Langergraber, G., Muellegger, E., 2005. Ecological Sanitation-a way to solve global sanitation problems? Environment International 31, 433-444.

Legros, D., Mccormick, M., Mugero, C., Skinnider, M., Bek'obita, D.D. & Okware S.I., 2000 Epidemiology of cholera outbreak in Kampala, Uganda. East African Medical Journal 77:347-349.

Mara, D., Lane, J., Scott, B., Trouba D., 2010. Sanitation and Health. PLoS Medicine 7, 1-7.

Meleg, E., Jakab, F., Kocsis, B., Bányai, K., Melegh, B., Szucs, G., 2006. Human astroviruses in raw sewage samples in Hungary. Applied and Environmental Microbiology 101, 1123–1129.

Mena, K.D., Gerba, C.P., 2009 Waterborne adenovirus. Rev Environ Contam Toxicol 198, 133-67.

Miagostovich, M.P., Ferreira, F.F.M., Guimarães, F.R., Fumian, T.M., Diniz-Mendes, L., Luz, S.L.B., Silva, L.A., Leite, J.P.G., 2008. Molecular Detection and Characterization of Gastroenteritis Viruses Occurring Naturally in the Stream Waters of Manaus, Central Amazônia, Brazil. Applied and Environmental Microbiology 74, 375–382.

Moe, C.L., Rheingans, R.D., 2006. Global challenges in water, sanitation and health. J Water Health 04 Supplement 1, 41-57.

Montgomery, M.A., Elimelech, M., 2007. Water And Sanitation in Developing Countries: Including Health in the Equation. Environ. Sci. Technol. 41, 17–24.

Moore R.D. 2003. Introduction to Salt Dilution Gauging for Streamflow Measurement: Part 1. Streamline Watershed Management Bulletin 7, 20-23.

Muscillo, M., Pourshaban, M., Iaconelli, M., Fontana, S., DiGrazia, A., Manzara, S., Fadda, G., Santangelo, R., 2008. Detection and quantification of human adenoviruses in surface waters by nested PCR, TaqMan real-time PCR and cell culture assay. Water Air Soil Poll 191, 83–93.

Noble, R.T., Fuhrman, J. A., 1997. Virus Decay and Its Causes in Coastal Waters. Applied and Environmental Microbiology 63, 77–83.

Oswald, W.E., Lescano, A.G., Bern, C., Calderon, M.M., Cabrera, L., Gilman, R.H., 2007. Fecal Contamination of Drinking Water within Peri-Urban Households, Lima, Peru. Am. J. Trop. Med. Hyg 77, 699–704.

Pang, X.L., Lee, B., Boroumand, N., Leblanc, B., Preiksaitis, J.K., Yu Ip, C.C., 2004. Increased detection of rotavirus using a real time reverse transcription polymerase chain reaction (RT PCR) assay in stool specimens from children with diarrhea. Journal of Medical Virology 72, 496-501.

Paterson, C., Mara, D., Cutis, T., 2007. Pro-poor sanitation technologies. Geoforum 38, 901-907.

Pedley, S., Yates, M., Schijven, J.F., West, J., Howard, G., Barrett M., 2006. Pathogens: Health, transport and attenuation ed. Schmoll, O., G. Howard, G., Chilton, J. and Chorus, I. PP 40-80. London: World Health Organization.

Pina, S., Puig, M., Lucena, F., Jofre, J., Girones, R., 1998. Viral pollution in the environment and in shellfish: human adenovirus detection by PCR as an index of human viruses. Applied and Environmental Microbiology 64, 3376–3382.

Ramani, S., Kang, G., 2009. Viruses causing childhood diarrhoea in the developing world. Curr Opin Infect Dis. 22, 477-82.

Rigotto, C., Hanley, K., Rochelle, P.A., De Leon, R., Barardi, C.R.M., Yates M.V., 2011. Survival of Adenovirus Types 2 and 41 in Surface and Ground Waters Measured by a Plaque Assay. Environ. Sci. Technol 45, 4145–4150.

Rigotto, C., Victoria, M., Moresco V., Kolesnikovas, C.K., Corrêa, A.A, Souza, D.S.M., Miagostovich, M.P., Simões, C.M.O., Barardi, C.R.M., 2010. Assessment of adenovirus, hepatitis A virus and rotavirus presence in environmental samples in Florianopolis, South Brazil. Journal of Applied Microbiology 109, 1979–1987.

Rodríguez, R.A., Pepper, I.L., Gerba, C.P., 2009. Application of PCR-Based Methods To Assess the Infectivity ofEnteric Viruses in Environmental Samples. Applied and Environmental Microbiology 75, 297–307.

Schvoerer, E., Bonnet, F., Dubois, V., Cazaux, G., Serceau, R., Fleury, H.J.A., Lafon, M.E., 2000. PCR detection of human enteric viruses in bathing areas, waste waters, and human stools in southwestern France. Research in microbiology 151, 693-701.

Templeton, M.R., Andrews, R.C., Hofmann, R., 2008. Particle-Associated Viruses in Water: Impacts on Disinfection Processes, Critical Reviews in Environmental Science and Technology 38, 137-164.

van Cuyck, H., Fan, J., David L. Robertson, D.L., Roques, P., 2005. Evidence of Recombination between Divergent Hepatitis E Viruses. Journal of Virology 79, 9306–9314.

van Geen, A., Ahmed, K.M, Akita, Y., Alam, Md. J., Culligan, P.J., Emch, M., Escamilla, V., Feighery, J., Ferguson, A.S., Knappett, P., Layton, A.C., Mailloux, B.J., McKay, L.D., Mey, J.L., Serre, M.L., Streatfield, P.K., Wu, J., Yunus, M., 2011. Fecal

Contamination of Shallow Tubewells in Bangladesh Inversely Related to Arsenic. Environ. Sci. Technol 45, 1199-1205.

van Heerden, J., Ehlers, M.M., Heim, A., Grabow, W.O.K., 2005. Prevalence, quantification and typing of adenoviruses detected in river and treated drinking water in South Africa. Journal of Applied Microbiology 99, 234-242.

van Zyl, W.B., Page, N.A., Grabow, W.O.K., Steele, A.D., Taylor, M.B., 2006. Molecular epidemiology of group A rotaviruses in water sources and selected raw vegetables in southern Africa. Applied and Environmental Microbiology 72, 4554–4560.

Vasickova P., Pavlik, I., Verani, M., Carducci, A., 2010. Issues Concerning Survival of Viruses on Surfaces Food and Environmental Virology 2, 24-34.

Venter, J. M. E., van Heerden, J., Vivier, J. C., Grabow, W.O.K., Taylor, M.B., 2007. Hepatitis A virus in surface water in South Africa: what are the risks? Journal of Water and Health 5, 229-239.

Verheyen, J., Timmen-Wego, M., Laudien, R., Boussaad, I., Sen S., Koc A., Uesbeck A., Mazou F., Pfister, H., 2009. Detection of Adenoviruses and Rotaviruses in Drinking Water Sources Used In Rural Areas of Benin, West Africa. Applied and Environmental Microbiology 75, 2798–2801.

Vilaginès, Ph., Sarrette, B., Husson, G. & Vilaginès, R., 1993. Glass Wool for Virus Concentration at Ambient Water pH Level. Water Science and Technology 27, 299–306.

WHO and UNICEF, 2010. *Progress on drinking water and sanitation*. Joint Monitoring Program Report (JMP). Geneva, Switzerland: World Health Organisation.

Wolfaardt, M., Moe, C.L., Grabow, W.O.K., 1995. Detection of small round structured viruses in clinical and environmental samples by polymerase chain reaction. Water Science and Technology 31, 375–382.

Wyn-Jones, A.P., Carducci, A., Cook, N., D'Agostino, M., Divizia, M., Fleischer, J., Gantzer, C., Gawler, A., Girones, R., Höller, C., de Roda Husman, A.M., Kay, D., Kozyra, I., López-Pila, J., Muscillo, M., Nascimento, M.S.J., Papageorgiou, G., Rutjes, S., Sellwood, J., Szewzyk, R., Wyer, M., 2011. Surveillance of adenoviruses and noroviruses in European recreational waters. Water Research 45, 1025-1038.

Wyn-Jones, A.P., Sellwood, J., 2001. Enteric viruses in the aquatic environment. Journal of Applied Microbiology 91, 945-962.

Xagoraraki, I., Kuo, D.H.W., Wong, K., Wong, M., Rose, J.B., 2007. Occurrence of human Adenoviruses at two recreational beaches of the Great Lakes. Applied and Environmental Microbiology 73, 7874–7881.

Chapter 5: Quantification of microbial risks to human health caused by waterborne viruses and bacteria in an urban slum

This Chapter is based on:

Katukiza, A.Y., Ronteltap, M., van der Steen, J.W., Foppen, J.W.A., Lens, P.N.L., 2013. Quantification of microbial risks to human health caused by waterborne viruses and bacteria in an urban slum. *Journal of Applied Microbiology*, DOI: 10.1111/jam.12368. "In Press".

Abstract

Microbial risks to human health in urban slums are caused by enteric viruses, bacteria, parasitic protozoa and helminths from excreta discharged into the slum. The magnitude of these risks based on the measured concentration of pathogens, especially waterborne viruses, through various exposure pathways in urban slums is not known. A quantitative microbial risk assessment (QMRA) was carried out to determine the magnitude of microbial risks from waterborne pathogens through various exposure pathways in Bwaise III in Kampala (Uganda). This was based on the concentration of *E. coli* O157:H7, *Salmonella* spp., rotavirus (RV) and human adenoviruses F and G (HAdv) in spring water, tap water, surface water, grey water and contaminated soil samples. The total disease burden was 680 disability-adjusted life years (DALYs) per 1000 persons per year. The highest disease burden contribution was caused by exposure to surface water open drainage channels (39%) followed by exposure to grey water in tertiary drains (24%), storage containers (22%), unprotected springs (8%), contaminated soil (7%) and tap water (0.02%). The highest percentage of the mean estimated infections was caused by *E. coli* O157:H7 (41%) followed by HAdv (32%), RV (20%), and *Salmonella* spp. (7%). In addition, the highest infection risk was 1 caused by HAdv in surface water at the slum outlet, while the lowest infection risk was 2.71×10^{-6} caused by *E. coli* O157:H7 in tap water. The results show that the slum environment is polluted and the disease burden from each of the exposure routes in Bwaise III slum, with the exception of tap water, was much higher than the WHO reference level of tolerable risk of 1×10^{-6} DALYs per person per year. The findings this study provide guidance to governments, local authorities and non-government organisations in making decisions on measures to reduce infection risk and the disease burden by 10^2 to 10^5 depending on the source of exposure to achieve the desired health impacts. The infection risk may be reduced by sustainable management of human excreta and grey water, coupled with risk communication during hygiene awareness campaigns at household and community level. The data also provide a basis to make strategic investments to improve sanitary conditions in urban slums.

5.1   Introduction

Microbial risks to human health are caused by enteric viruses, bacteria, protozoa and helminths as result of poor sanitation and hygiene practices (Ashbolt, 2004; Montgomery and Elimelech, 2007; Buttenheim, 2008; Mara et al., 2010). The risk of infection from pathogenic microorganisms depends on die-off rates, attenuation and dilution factors (Alexander et al., 1986; Dowd et al., 2000; Ferguson et al., 2003; Pedley et al., 2006). These factors determine the concentration of pathogens in the urban slum, which is defined as an informal settlement in a city or town characterised by poor urban infrastructure, low water and sanitation service levels, high population density and limited access for basic services.

The potential exposure routes from contaminated potable water, wastewater, soils and food sources include intentional and accidental ingestion, dermal contact and inhalation (Stanek and Calabrese, 1995; Haas et al., 1999; Howard et al., 2003; Styen et al., 2004; Westrell et al., 2004; Schönning et al., 2007). In slums, the incidence of diarrhoea, dysentery and gastroenteritis is especially high in immuno-compromised groups that include children, the elderly and pregnant women and is attributed to person-to-person contact as well as faecally contaminated water and soil (Alirol et al., 2010; Victora et al., 1988; Muoki et al., 2008). Therefore, prevention of contact between untreated excreta and the environment in slums reduces the risk of infections since human and animal faeces are the primary sources of pathogens.

The public health risks from sanitation systems are a result of inadequate collection, storage, treatment, disposal or re-use of excreta, grey water and solid waste (Cronin et al., 2009; Martens and Böhm, 2009; Ottoson and Stenström, 2003; Peterson et al., 2011; Schönning et al., 2007; Stenström et al., 2011). These risks may be higher in urban slums because of the lack of adequate sanitation provision (Katukiza et al., 2012). Infections related to poor sanitation and hygiene practices remain the leading cause of mortality and morbidity in urban slums (Alirol et al., 2010; Bezerra et al., 2011; Butala et al., 2010; Buttenheim, 2008; Mara et al., 2010; WHO and UNICEF, 2012), despite sanitation interventions providing sanitation facilities (Joyce et al., 2010; Lüthi et al., 2010; Mara and Alabaster, 2008). The situation is made worse by further stress on the water supply and sanitation infrastructure as a result of the high urbanization rates, which are comparable to the growth rate of urban slums in developing countries (Isunju et al., 2011; Ooi and Phua, 2007).

The magnitude of microbial risks to human health from pathogens, especially waterborne viruses in typical urban slums in developing countries remains unknown. Therefore, also the effects of efforts to improve slum sanitation on the health of slum dwellers remain unknown. Moreover, in the very few Quantitative Microbial Risk Assessment (QMRA) studies that were carried out in the urban poor areas (Howard et al., 2006b; Labite et al., 2010; Machdar et al., 2013), the *E. coli* to pathogen ratios from the literature were used instead of the measured concentration of waterborne viruses, which may have increased variability and uncertainty in the estimated risk of infection. In addition, there is limited information on the quantification of risks from exposure to grey water, which is the largest volumetric wastewater flux in urban slums.

This study was conducted in a typical slum in Kampala (Uganda) with very poor sanitation infrastructure. Our main objective was to determine the magnitude of microbial risks to human health caused by waterborne viruses and bacteria in the urban slum. The specific objectives were to determine the contribution of the sources of contamination to the disease burden, the proportion of infections caused by each reference pathogen and to evaluate the health effects of possible interventions to reduce the risk of infections.

5.2   Materials and methods

5.2.1   The study area

The study area of Bwaise III in Kampala city (Uganda) is an informal settlement located in a reclaimed wetland with high water table (32° 34'E and 0° 21'N). It is drained by two major open surface water drainage channels to which tertiary drains convey grey water discharge. The area has poor sanitation infrastructure, a high population density of over 250 persons per hectare and limited accessibility to basic services, which are characteristics of a typical slum area in Sub-Saharan Africa. The area has a high population growth rate estimated by local authorities to be 9% per year. Children under 5 years account for 21% of the population (UBOS 2011). The existing management systems for excreta, grey water and solid waste are inadequate (Katukiza et al., 2010). The main diseases reported at the health centre in the area are malaria, diarrhoea, dysentery and cholera with the highest prevalence occurring in children below 5 years. In Bwaise III, more than 80% of the residents are not formally employed and engage in small-scale businesses like general merchandising and food vending. The individual monthly income ranges from US$ 10 to US$ 250. There is piped water supply infrastructure but most residents prefer to obtain free water from the existing spring water sources, which are contaminated (Katukiza et al. 2013).

5.2.2   Hazard Identification

Pathogens were chosen for the QMRA based on the following criteria: occurrence and persistence in the slum environment; low infectious doses; possibilities for detection and quantification; adequate literature on the organism; representativity of a major group of pathogens; and the occurrence of diseases such as diarrhoea and gastroenteritis in the population reported by the health authorities in the study area. *E. coli*, *Salmonella* spp. and total coliforms were chosen to represent bacteria, while rotavirus and human adenoviruses F and G were chosen to represent the viruses.

These were the most relevant pathogens and indicator organisms in the study area and were used as reference pathogens for the QMRA. Protozoa and helminths were not included in the QMRA because of sample analysis limitations.

5.2.3 Exposure Assessment

The potential exposure routes in the study area were identified to determine the critical points to quantify the microbial risks to human health. The hazard pathways identified included ingestion, dermal contact and inhalation. Sampling locations for surface water, grey water, water sources and soil samples were identified in each of the zones. This was based on the location and number of main storm water drains, grey water tertiary drains, water sources and playing grounds for children. The likely exposed population was then determined in consultation with the local environmental and health officers. In this study, we assumed that the microbial risks from dermal contact and inhalation were minor relative to exposure through ingestion. In addition, exposure to contaminated food sources was not considered in our study because the residents obtain food from outside the study area. The volume of waste water or water ingested (mL per day) or weight of soil ingested (g per day) was multiplied by the pathogen concentration in a sample to calculate the total exposure or dose for a particular sampling point. The concentrations of pathogens and indicator organisms were determined as described under the sample collection and analysis procedure.

Table 5.1 shows the sources of contamination and underlying assumptions that were used in the exposure assessment for the representative organisms. Figure 5.1 shows a typical exposure to pathogens scenario of children in Bwaise III through dermal contact and ingestion of contaminated surface water and spring water. The proportion of the population using unprotected springs was 11.8% and tap water was 87.8% (0.3% for house connections, 74.5% for public yard taps and 13 for yard taps; Katukiza et al., 2010).

Table 5.1: Exposure routes, sources of contamination and assumptions for estimation of pathogen dose

Exposure route	Representative organisms	Source of contamination	Volume ingested and underlying assumptions	References
Ingestion				
	E.coli, Salmonella spp., Total Coliforms, Adenoviruses (F&G), Rotavirus.	Drinking Water	Exposure is based on the assumption of drinking of 1-2 L per person per day. For a slum area 500-800 mL is reasonable. Ingestion of 500ml per person per day was used. It was assumed that 40% of the total population drink water stored in containers at household level and that tap water from the piped water supply system is available for 5 days in a week.	Howard et al., 2006; steyn et al., 2004.
	E.coli, Salmonella spp., Total Coliforms, Adenoviruses (F&G), Rotavirus.	Grey water	Exposure is through direct contact by kids playing in the tertiary drains conveying grey water, body washing, preparing of food and eating from near the tertiary drains in slums. Involuntary ingestion of 1-5 mL and unintentional immersion during flooding 10-30 mL. Frequency of exposure to 5 mL grey water of 6 per year was used.	Labite et al., 2010; Steyn et al., 2004; Westrell et al., 2004.
	E.coli, Salmonella spp., Total Coliforms, Adenoviruses (F&G), Rotavirus.	Surface Water	Exposure through direct contact by kids playing and swimming in the open channels, body washing, exposure during flooding of the slum area. Involuntary ingestion of 1-5 mL and unintentional immersion during flooding 10-30 mL. The ingestion of 10 mL adopted was based on a frequency of 6 per year	Labite et al., 2010; Steyn et al., 2004; Westrell et al., 2004.
	E.coli, Salmonella spp., Total Coliforms	Soil	Exposure range of 1-10 gram per person per day. The frequency of 12 per year from 10 g of contaminated soil to to children under 5 years who account for 21% of the population.	Mara et al., 2007; UBOS, 2011; Westrell et al., 2004.
	Viruses and bacteria	Food*	Food consumed in the study area, Bwaise III originates from other areas and this exposure route was not considered.	
Dermal	Adenovirus (F&G) Rotavirus	Surface water and grey water	Exposure through direct contact by kids playing in the open channels and tertiary drains, contact with sands by construction laborers excavating sand in the drains, body washing and swimming at downstream side of the open channel.	Mena and Gerba, 2009
	E.coli, Salmonella spp., Total Coliforms,	Soil	Contact with contaminated soil.	
Inhalation	Viruses, bacteria	Aerosols*	The microbial risks associated with inhalation bacteria and virus particles are minor relative to those from other exposure routes.	

* Sources of contamination in the study area that were not considered in the QMRA

111

A   B

Figure 5.1:   **(A) Children collecting and drinking water from a polluted stream in Bwaise III, and (B) Collection and drinking of raw spring water in the slum**

5.2.4   Sample collection and analysis for detection of bacteria

A total of 633 samples were collected in sterilised containers from 54 sampling locations in the study area in the morning hours (between 8:30 am and 10 am) during the period January 04, 2010 and February 3, 2011. The sampling locations included: 11 for grey water, 26 for surface water, 5 for contaminated soil, 5 for tap water, 2 for spring water and 5 for in-house vessels or storage containers. The selection criteria of the sampling points included: outlets and inlets of major open storm water drainage channels, tertiary drains conveying grey water directly from bathrooms, spring water samples from households that obtain water from existing spring sources only, and open space that act as playing grounds for children where sand excavated from drains is dumped.

For soil sampling, a soil auger was used to obtain samples from the top 15 cm layer three times from the same location, which were homogeneously mixed to get a representative sample. 10 g of the homogeneous sample was added to 100 mL of dilution water (8.5g NaCl and 1g Peptone dissolved in 1 litre of distilled water to provide optimum conditions for survival of *E. coli*, *Salmonella* spp. and total coliforms), shaken for 30 minutes using a dual action shaker (Model no. 3508-1, Lab-line Instruments Inc.) and then allowed to stand for 1 hour for the particles to settle. The supernatant was then analysed using the same standard methods as for the water samples.

After collection, the samples were stored in the dark at +4°C using ice blocks and transported to the Public Health Engineering Laboratory at Makerere University (Uganda) for processing within one hour. The concentration of *Escherichia coli*, Total coliforms and *Salmonella* spp. was determined with Chromocult® Coliform Agar media using the spread plate method (APHA, AWWA and WEF, 2005).

5.2.5   Sample collection and analysis for detection of waterborne viruses

A total of 40 samples were collected from storm water drainage channels, grey water tertiary drains and unprotected springs during the period January 04, 2011 and February 3, 2011 (Katukiza et al., 2013). The glass wool filtration protocol described by Wyn-Jones et al. (2011) (a modification of the method of Vilaginès et al. (1993)) was used to concentrate virus particles from 10 L samples. Viral concentrates were kept frozen during transportation from Uganda to the laboratories of UNESCO-IHE (Delft, The Netherlands), where they were immediately stored at -80°C until further analysis. Sterilised pipettes, filter tips, vials and reagent tubes were used during sample processing to avoid contamination of samples and reagents. Viral nucleic acids were then extracted by a procedure described by Boom (Boom et al., 1990; Boom et al., 1999).

The detection of human adenoviruses F and G (HAdv) was carried out using USB HotStart-IT Probe qPCR Master Mix (USB Cooperation, Cleveland, OH, USA) and virus particles were quantified using a standard curve constructed in triplicate from a dilution series of a concentrated genomic copy number standard (2×10^5 copies μL^{-1}), that was included in the standard kit for quantification of human adenovirus type F and G (from Primerdesign). The detection of rota viruses (RV) was carried out using the Platinum Quantitative RT-PCR ThermoScript One-Step System (Invitrogen Life Technologies Cooperation, Grand Island, NY, USA). RV concentrations were determined with a standard curve, which was constructed in duplicate from a dilution series of concentrated genomic copy synthetic cDNA stock (2.41×10^7 copies μL^{-1}) with a length of 87 base pairs (position 963-1049; Pang et al., 2004). The details of primers and probe sequences used for detection of HAdv and RV and the corresponding standard curves were as described by Katukiza et al. (2013).

5.2.6   Dose-response models

The dose-response models were used to estimate the probability of an infection that occurs depending on the incubation period (Haas et al., 1999; Mayer et al., 2011; Mena and Gerba, 2009). The dose-response models used were the β-Poisson model for rotavirus infections (Haas et al., 1999: α = 0.2531 and N_{50} = 6.17), E. coli O157:H7 (Haas et al., 2000: α = 0.49 and N_{50} = 596,000), Salmonella spp. (Haas et al., 1999; Styen et al., 2004: α = 0.3126 and N_{50} = 23,600) and the exponential model for Adeno virus (Crabtree et al., 1997; Mena and Gerba, 2009: r =0.4172). The exponential model was based on the assumption that organisms have independent and identical probability of surviving to reach an appropriate site and cause an infection while the Beta-Poisson model took into account infection and survival probabilities of the host (Haas et al., 1999; Mena and Gerba, 2009). The following model equations were used:

(a) β-Poisson dose-response model

$$P_I(d) = 1 - \left[1 + \left(\frac{d}{N_{50}}\right)\left(2^{1/\alpha} - 1\right)\right]^{-\alpha}$$ ...... (1)

(b) Exponential dose-response model

$$P_I(d) = 1 - \exp\left(-rd\right)$$ ...... (2)

(c) Annual risk of infection

$$P_{I(A)}(d) = 1 - \left[1 - P_I(d)\right]^n$$ ...... (3)

Where $P_I(d)$ is the probability or risk of infection for an individual exposed to a single pathogen dose d through ingestion or dermal contact; α and r are parameters that characterise dose-response relationships referred to as pathogen infectivity constants; d is the pathogen dose; N_{50} is the median infective dose or the number of pathogens required to cause an infection in 50% of the exposed population. $P_{I\ (A)}(d)$ is the estimated annual probability or risk of an infection from n exposures per year due to a single pathogen dose d.

5.2.7 Risk characterisation

Risk characterisation was done to integrate hazard identification, exposure assessment and the dose-response relationship to determine a health outcome (risk of infection, illness and mortality). The possible adverse health effects to humans from exposure to pathogens were represented by Disability Adjusted Life Years (DALYs) (Fewtrell et al., 2007; Kemmeren et al., 2006; Lopez et al., 2006). The following equation was used for the calculation of DALYs per year for each case:

$$DALY = \sum_{n=1}^{n} P_{ill/inf} \times P_{outcome\ i/ill} \times D_i \times S_i$$ ...... (4)

Where n is the total number of outcomes considered, $P_{(ill/inf)}$ is the probability of illness given infection, $P_{(outcome/ill)}$ is the probability of outcome i given illness, D_i is the duration (years) of outcome i and S_i is the severity weight for outcome i.

The measured concentrations of HAdv and RV in genomic copies per mL was used in the QMRA as actual concentration of virus particles each capable of causing an infection because free nucleic acid (RNA or DNA) is unstable in the environment. This leads to an overestimation of the infection risk because the concentration of infectious particles used in determining the dose can be highly variable. In addition,

114

infectivity depends on whether the viral capsid is damaged or not for both enveloped and non-enveloped viruses (Templeton et al., 2008; Rodríguez et al., 2009). The measured concentration of *E. coli* was multiplied by a factor of 0.08 to obtain the concentration of the pathogenic strain of *E. coli* O157:153 H7 and *Salmonella* spp. for use in the QMRA (Haas et al., 1999; Machdar et al., 2013). The determination of pathogenic strains of *E. coli* and *Salmonella* spp. based on this factor gives an approximation of the actual concentration of the human pathogens, which may have increased the uncertainty in the estimated risk of infection.

5.2.8   Burden of the disease determination

The disease burden for each case was calculated as the product of the probability of illness outcome from an infection, the severity factor and the duration in years and was expressed in DALYs per year (Haas et al., 1999; Havelaar and Melse, 2003; Kemmeren et al., 2006; Mathers et al., 2006). The severity weight and duration of the disability or disease adjusted to conditions in developing countries were used (Table 5.2). The severity weights' scale ranges from 0 for healthy to 1 for death outcome. The average life expectancy at birth of 53.4 year was obtained from the national statistics for Uganda (UBOS, 2011). The population distribution in Bwaise III local council zones was used to determine the exposed population at each sampling point subject to the underlying assumptions as shown in Table 5.1. The percentage of the total Bwaise III population in the Kawaala, Katoogo, St. Francis, Bugalani, Bukasa and Kalimali local council I zones was 10%, 25%, 17.8%, 16%, 16.3% and 15% respectively.

Table 5.2:   The severity weight, duration and disease burden for each pathogen

Pathogen type	Pathogen	Relevant Health outcomes	Severity weight	Duration		Disease burden (DALYs per case)	References
				days	years		
Viruses							
	Rota virus	*Diarrhoea*					Havelaar and Melse, 2003; Howard et al., 2006b
		Mild diarrhoea; 85.6% of all cases	0.1	7	0.019	1.6×10^{-3}	
		Severe Diarrhea; 14.4% of all cases	0.23	7	0.019	1.6×10^{-4}	
		Death from diarrhoea; 0.7%	1	19144.25	52.45	3.7×10^{-1}	
		Total				**0.369**	
	Human adenoviruses, F and G	*Diarrhoea*					Mathers et al., 2006; Mena and Gerba, 2009
		Mild diarrhoea; 85.6% of all cases	0.1	3.4	0.009	7.6×10^{-4}	
		Severe Diarrhea; 14.4% of all cases	0.23	5.6	0.015	5.1×10^{-4}	
		Death from diarrhoea; 0.5%	1	19144.25	52.45	5.2×10^{-2}	
		Gastroenteritis					Mathers et al., 2006; Mena and Gerba, 2009
		Mild illness, 50%	0.099	5	0.014	6.8×10^{-4}	
		Probability of seeing GP, 5%	0.279	10	0.027	3.8×10^{-4}	
		Death from illness; 0.1%	1		52.45	5.2×10^{-2}	
		Total				**0.527**	
Bacteria							
	Enteropathogenic E. coli	*Gastroenteritis* Probability of not seeing a general practitioner (GP); 94% of all cases	0.067	5.58	0.015	9.6×10^{-4}	Havelaar and Melse, 2003.
		Probability of seeing GP; 6%	0.393	10.65	0.029	6.9×10^{-4}	
		Probability of being hospitalised; 9%	0.393	16.15	0.044	1.6×10^{-3}	
		Fatal; 0.2%	1	19144.25	52.45	1.6×10^{-3}	
		Diarrhoea					Havelaar and Melse, 2003
		Waterly diarrhoea; 53% of all cases	0.067	3.4	0.009		
		Bloody diarrhoea; 43% of cases	0.39	5.6	0.015		
		Death from Diarrhoea; 0.2%	1		52.45		
		Haemolytic uraemic syndrome (HUS); 1% of all cases	0.93	21	0.058		
		Death from HUS; 0.1% of all cases	1	19144.25	52.45		
		Total				**0.269**	
	Salmonella	*Gastroenteritis* Probability of not seeing a general practitioner (GP); 94% of all cases	0.067	5.58	0.0153		Haas et al. 1999; Kemmeren et al., 2006
		Probability of seeing GP; 6%	0.393	10.65	0.0292		
		Probability of being hospitalised; 9%	0.393	16.15	0.0442		
		Fatal; 0.1%	1	19144.25	52.45		
		Reactive arthritis (RA), 8% of patients with gastroenteritis develop RA					Kemmeren et al., 2006
		No visit to GP, 85.5%	0.127	222	0.6082		
		Visit to GP, 22%	0.21	222	0.6082		
		Hospitalised, 2.2%	0.37	222	0.6082		
		Total				**0.089**	

5.2.9   Variability and uncertainty in the data

The concentration of microorganisms (pathogens and indicator organisms) varied depending on the anthropogenic activities in Bwaise III. The extent of this variability was uncertain. In order to make the exposure estimates used in this study reliable, Monte Carlo simulations were made for 100,000 iterations using XLSim software®3. The log-normal distributed random variable was based on the mean (μ) and standard deviation (δ) of the log of the measured concentration of a pathogen. The probability of infection was calculated for each variable based on random selection of a value from the log-normal distributed probability density function (PDF). The PDF catered for the range and variation of concentration values with time. Figure 5.2 shows an example of how the log-normal distribution fitted the data based on a statistical test. The output of the analysis was the mean and standard error of the risk of infection as well as the frequency distribution of the probabilities of infection. The minimum concentration of pathogens was set at zero to avoid the risk output with negative values.

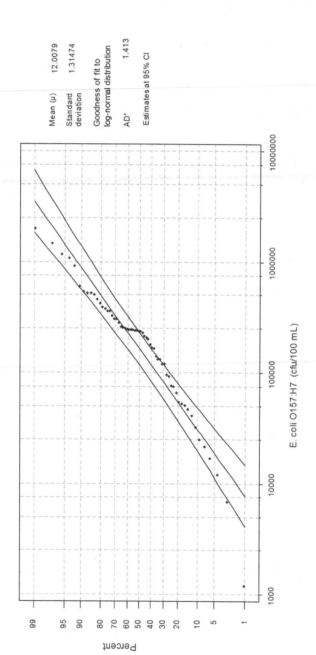

Mean (μ)   12.0079
Standard   1.31474
deviation

Goodness of fit to
log-normal distribution

AD*   1.413

Estimates at 95% CI

Figure 5.2: **A probability plot showing how the concentration of *E. coli* O157:H7 in samples obtained from Nsooba channel inlet fit the log-normal distribution using Anderson–Darling (AD) statistical test**

5.3 Results

5.3.1 Sources of contamination and concentrations of bacteria and waterborne viruses

The sources of contamination identified in the QMRA were open drainage channels, grey water tertiary drains, contaminated soil, spring water sources and in-house storage vessels or containers. Table 5.3 shows the concentration of *E. coli*, *Salmonella* spp. and total coliforms in 633 samples from 54 sampling locations. The highest concentrations in samples that tested positive (n= 562) amounted to $8.01 \times 10^6 (\pm 8.15 \times 10^6)$ cfu.$(100 \text{ mL})^{-1}$ for *E. coli* in grey water, $1.9 \times 10^5 (\pm 2.32 \times 10^5)$ cfu.$(100 \text{ mL})^{-1}$ for *Salmonella* spp. in surface water and $5.11 \times 10^7 (\pm 2.32 \times 10^5)$ cfu.$(100 \text{ mL})^{-1}$ for total coliforms in grey water. The lowest concentrations in samples that tested positive (n=562) amounted to $3.33 \times 10^{-1} (\pm 5.1 \times 10^{-1})$ cfu.$(100 \text{ mL})^{-1}$ for *E. coli* in tap water, $6.67 \times 10^0 (\pm 2.58 \times 10^1)$ cfu.$(100 \text{ mL})^{-1}$ for *Salmonella* spp. in water from storage containers and $7.78 \times 10^{-1} (\pm 1.2 \times 10^0)$ cfu.$(100 \text{ mL})^{-1}$ for total coliforms in tap water. The average ratio of the concentration (Log_{10}) of *E. coli* to total coliforms was 0.9 for surface water samples and 0.6 for grey water samples while the corresponding ratio of *E. coli* to *Salmonella* spp. was 1.3 and 3.1.

The concentration of HAdv and RV in 26 samples (65% of the total number of samples) that tested positive was expressed in genomic copies per mL (gc.mL^{-1}) (Table 5.4). The highest concentration of HAdv was 2.65×10^1 gc.mL^{-1} in surface water at the outlet of the main drainage channel, while the lowest was 7.62×10^{-3} gc.mL^{-1} in a spring water sample. The highest concentration of RV was 5.12 gc.mL^{-1} in surface water at the outlet of the main drainage, while the lowest was 2.96×10^{-1} gc.mL^{-1} in a surface water sample. The average *E. coli*:HAdv ratio was $8.3 \times 10^6 \pm 1 \times 10^7$ for surface water samples and $1.2 \times 10^7 \pm 1.5 \times 10^6$ for grey water samples, while the average *E. coli*:RV ratio was $4.5 \times 10^7 \pm 6.1 \times 10^7$ for surface water samples and $2.9 \times 10^6 \pm 2.8 \times 10^6$ for grey water samples. The *E. coli* to HAdv ratio for spring water was 3.9×10^3:1. The detection of waterborne viruses in 65% of all surface water and grey water samples was an indication of inadequate human excreta management in the urban slum. The sample size used in this study for detection of waterborne viruses was limited by transportation of materials and samples for analysis from Makerere University in Uganda to UNESCO-IHE laboratories in Delft (The Netherlands), which may have increased uncertainty in the mean estimated infection rates.

Table 5.3: Concentration of *Escherichia coli*, *Salmonella* spp. and total coliforms in samples from different sources of contamination*

Source of contamination	Sampling location	Bacteria	Concentration (log$_{10}$ cfu/100ml)		Number of samples
			Mean	Standard deviation	
Open storm drainage channels	Nsooba Inlet	*Escherichia coli*	6.6	6.4	60
		Salmonella spp.	4.6	4.5	
	Nsooba outlet	*Escherichia coli*	6.6	6.3	60
		Salmonella spp.	4.6	4.4	
	JN Nsooba/Nakamilo	*Escherichia coli*	6.5	6.2	60
		Salmonella spp.	5.1	5.2	
	Nakamilo inlet	*Escherichia coli*	6.9	6.9	60
		Salmonella spp.	5.3	5.4	
Open grey water tertiary drains	Grey water in tertiary drain (st. Francis Zone)	*Escherichia coli*	6.2	6.2	60
		Salmonella spp.	5.0	4.9	
	Grey water tertiary drain (Katogo zone)	*Escherichia coli*	6.9	6.9	60
		Salmonella spp.	5.1	5.2	
	Grey water tertiary drain (Bombo road)	*Escherichia coli*	6.4	6.7	60
		Salmonella spp.	4.7	4.5	
	Grey water tertiary drain (Kawaala road)	*Escherichia coli*	6.1	6.4	60
		Salmonella spp.	4.6	4.5	
Contaminated soil	Soil from open space; play ground for children	*Escherichia coli*	5.4	5.1	48
		Salmonella spp.	4.7	4.8	
Unprotected springs	Spring water from two sources[†]	*Escherichia coli*	2.5	2.9	30
		Salmonella spp.	ND	ND	
Tap water from pipe water supply system	5 yard taps and 5 public stand pipes	*Escherichia coli*	-0.5	-0.3	30
		Salmonella spp.	ND	ND	
Containers used to draw and store spring water	Vessels (drinking water collection containers) from 5 housholds	*Escherichia coli*	3.1	3.5	45
		Salmonella spp.	0.8	1.4	

*Previous studies: (A) *E. coli* (cfu/100mL): 10^2-10^6 in surface water (Byamukama et al., 2000), 10^2-10^4 in grey water (Kulabako et al., 2011), 10^3-10^5 in grey water (Carden *et al.*, 2007), 10^{-1} to 3.8x10^1 in drinking water sources (Machdar et al., 2013) (B): Total coliforms (cfu/100mL): 10^4-10^7 in surface water (Byamukama et al., 2000) 10^4-10^8 in wastewater from open drains and beach sand (Labite et al., 2010) (C) *Salmonella* spp.(cfu/100mL): 10^4-10^7 in wastewater from open drains and beach sand (Labite et al., 2010)

[†]There was no detection of *E. coli*, *Salmonella* spp. and total coliforms in samples from spring 2. ND refers to not detected

Table 5.4: Concentration of Human Adenoviruses (F&G) and Rotavirus in samples from different sources of contamination (Katukiza et al., 2013)

Source of contamination	Sampling location	Concentration (Genomic copies ml^{-1}) (N=40)	
		Human adenovirus (F&G)	Rotavirus
Open storm water drainage chan	Nsooba Inlet	1.53 (±1.1)	2.98x10^1 (±3.66X10^1)
	Nsooba outlet	2.65x10^1 (±1.9x10^1)	5.12 (±6.2)
	JN Nsooba/Nakamilo	5.32x10^{-1} (±4.0x10^{-2})	2.48(±9.61x10^{-2})
	Nakamilo inlet	3.27x10^{-1} (±4.8x10^{-2})	1.66(±5.63x10^{-1})
Open grey water tertiary drains	Grey water in tertiary drain (st. Francis Zone)	1.35x10^{-1} (±1.9x10^{-1})	3.44X10^{-1}(4.86X10^{-1})
	Grey water tertiary drain (Katogo zone)	7.80x10^{-1} (±1.6)	8.85 (±16.3)
	Grey water tertiary drain- Bombo road	ND	ND
	Grey water tertiary drain-Kawaala road	ND	ND
Unprotected springs	Spring water source 1	7.62x10^{-3} (±1x10^{-2})	ND
	Spring water source 2	ND	ND

In brackets is the standard deviation and ND refers to not detected

5.3.2 Risk of infection

The risk of infection from pathogenic bacteria and waterborne diseases was estimated for all sources of contamination (Table 5.5). The dose used in the determination of the risk of infections from the *E. coli* and *Salmonella* spp. was based on the volume or quantity ingested and the concentrations of the pathogenic strain. The highest health risk in Bwaise III was from HAdv at the outlet of the main open drainage channel (Nsooba) with a probability of infection of 1, while the lowest health risk was constituted by 8% (the estimated proportion of pathogen strain of *E. coli*: Haas et al., 1999; Machdar et al., 2013) of *E. coli* in tap water with a probability of infection of 2.7×10^{-6} (Table 5.5). The highest annual risk of infection from grey water exposure was 1 from HAdv and RV in Katoogo zone while the lowest was 3.64×10^{-1} from *Salmonella* spp. in the Kawaala road zone. The risk of infection from exposure to grey water in tertiary drains and contaminated surface water in open drainage channels was generally in the same order of magnitude (Table 5.5). The highest percentage of infections was 41% caused by *E. coli* followed by 32% of infections caused by HAdv (Figure 5.3).

Table 5.5: Quantity ingested, frequency, exposed population and the estimated risk of infection

Source of contamination	Sampling location	Microorganism	Quantity ingested (mL)*	Exposed population	Number of exposures to a single dose per	Dose; d	Probability of infection; P_I (d)	Annual probability of infection; $P_{I(A)}$ (d)
	Nsooba Inlet	Escherichia coli	10	676	6	3.18×10^4	1.43×10^{-1}	6.05×10^{-1}
		Salmonella spp.	10	676	6	3.56×10^2	9.39×10^{-2}	4.47×10^{-1}
		Adenoviruses (F&G)	10	676	6	1.53×10^1	9.98×10^{-1}	1.00×10^0
		Rotavirus	10	676	6	2.98×10^2	8.09×10^{-1}	1.00×10^0
	Nsooba outlet	Escherichia coli	10	901	6	3.01×10^4	1.29×10^{-1}	5.64×10^{-1}
		Salmonella spp.	10	901	6	3.09×10^2	8.28×10^{-2}	4.04×10^{-1}
		Adenoviruses (F&G)	10	901	6	2.65×10^2	1.00×10^0	1.00×10^0
Open storm drainage channels		Rotavirus	10	901	6	5.12×10^1	7.03×10^{-1}	9.99×10^{-1}
	JN Nsooba/Nakamilo	Escherichia coli	10	968	6	2.61×10^4	1.18×10^{-1}	5.30×10^{-1}
		Salmonella spp.	10	968	6	9.40×10^2	1.90×10^{-1}	7.17×10^{-1}
		Adenoviruses (F&G)	10	968	6	2.66×10^0	6.70×10^{-1}	9.99×10^{-1}
		Rotavirus	10	968	6	2.48×10^1	6.44×10^{-1}	9.98×10^{-1}
	Nakamilo inlet	Escherichia coli	10	968	6	6.03×10^4	2.42×10^{-1}	8.11×10^{-1}
		Salmonella spp.	10	968	6	1.52×10^3	2.17×10^{-1}	7.69×10^{-1}
		Adenoviruses (F&G)	10	968	6	3.27×10^0	7.44×10^{-1}	1.00×10^0
		Rotavirus	10	968	6	1.66×10^1	6.07×10^{-1}	9.96×10^{-1}
	Grey water in tertiary drain (st. Francis Zone)	Escherichia coli	5	800	8	6.76×10^3	6.41×10^{-2}	4.12×10^{-1}
		Salmonella spp.	5	800	8	4.42×10^2	1.13×10^{-1}	6.15×10^{-1}
		Adenoviruses (F&G)	5	800	8	5.75×10^{-1}	2.45×10^{-1}	8.95×10^{-1}
		Rotavirus	5	800	8	1.72×10^0	3.36×10^{-1}	9.62×10^{-1}
	Grey water tertiary drain (Katogo zone)	Escherichia coli	5	1,126	8	3.24×10^4	1.82×10^{-1}	7.99×10^{-1}
Open grey water tertiary drains		Salmonella spp.	5	1,126	8	4.56×10^2	1.36×10^{-1}	6.89×10^{-1}
		Adenoviruses (F&G)	5	1,126	8	3.90×10^0	8.03×10^{-1}	1.00×10^0
		Rotavirus	5	1,126	8	4.43×10^1	6.92×10^{-1}	1.00×10^0
	Grey water tertiary drain- Bombo road	Escherichia coli	5	732	8	1.08×10^4	1.43×10^{-1}	7.08×10^{-1}
		Salmonella spp.	5	732	8	2.03×10^2	5.94×10^{-2}	3.87×10^{-1}
		Adenoviruses (F&G)				ND		
		Rotavirus				ND		
	Grey water tertiary drain- Kawaala road	Escherichia coli	5	673	8	5.34×10^3	8.45×10^{-2}	5.06×10^{-1}
		Salmonella spp.	5	673	8	1.51×10^2	5.50×10^{-2}	3.64×10^{-1}
		Adenoviruses (F&G)				ND		
		Rotavirus				ND		
Contaminated soil	Soil from open space; play ground for children	Escherichia coli	5	3,153	12	9.14×10^3	6.38×10^{-2}	5.47×10^{-1}
		Salmonella spp.	5	3,153	12	2.24×10^3	2.43×10^{-1}	9.65×10^{-1}
		Adenoviruses (F&G)				ND		
		Rotavirus				ND		
Unprotected springs	Spring water from two sources	Escherichia coli	500	1,772	365	4.93×10^2	4.12×10^{-3}	7.78×10^{-1}
		Salmonella spp.				ND		
		Adenoviruses (F&G)	500	1,772	365	3.81×10^0	7.96×10^{-1}	1.00×10^0
		Rotavirus				ND		
Tap water from piped water supply sytem	Yard taps and public stand pipes	Escherichia coli	500	13,183	261	1.33×10^{-1}	2.71×10^{-6}	7.06×10^{-4}
		Salmonella spp.				ND		
		Adenoviruses (F&G)				ND		
		Rotavirus				ND		
Vessels (drinking water collection and storage containers)	Vessels (drinking water collection	Escherichia coli	500	6,006	365	1.27×10^3	1.58×10^{-2}	9.97×10^{-1}
		Salmonella spp.	500	6,006	365	2.67×10^0	2.89×10^{-4}	1.00×10^{-1}
		Adenoviruses (F&G)				ND		
		Rotavirus				ND		

*mL for liquid samples; g for soil and ND refers to microorganism not detected in the samples from that particular location

123

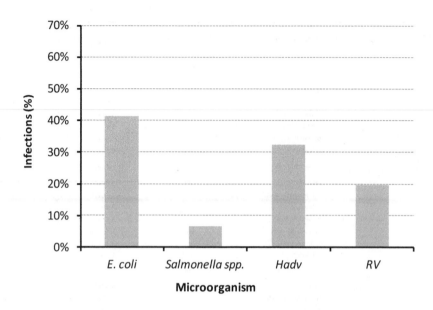

Figure 5.3:   **Percentage of infections caused by each microorganism (Hadv and RV refer to human adenoviruses F&G and rotaviruses, spectively)**

5.3.3 Disease burden

The disease burden in Bwaise III was estimated based on the health outcomes after exposure to RV, HAdv, enteropathogenic *E. coli* and *Salmonella* spp. (Table 5.5). The disease burden estimated for Bwaise III with an estimated population of 15,015 persons was 10,172 DALYs per year, which is equivalent to 680 DALYs per 1000 people per year (Table 5.6). The highest disease burden in the study area was in the Katoogo zone accounting for 29% of the total DALYs per year, while the lowest disease burden was in the Kawaala road zone accounting for 8% of the total DALYs per year (Figure 5.4).

Exposure from waterborne viruses and bacteria in open storm water drainage channels, grey water in tertiary drains and storage containers contributed 39%, 24% and 22% of the total DALYs per year in Bwaise III respectively. Exposure via tap water from a piped water supply system had the lowest disease burden contribution of 0.02% (Table 5.6).

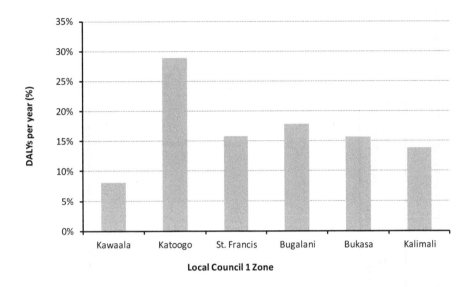

Figure 5.4:   **The distribution of the total disease burden (DALYs per year) in the zones of Bwaise III**

Table 5.6: The contribution of disease burden per year by different sources of contamination (the number of cases was calculated using Monte Carlo simulations)[*]

Source of contamination	Number of cases per year	Proportion of the total number of cases (%)	Disease burden (DALYs per year)	Disease burden (DALYs per person per year)	Proportion of the total disease burden (%)
Open storm drainage channels	11,340	34%	3,930	2.8×10^{-1}	39%
Open grey water tertiary drains	7,621	23%	2,392	1.3×10^{-1}	24%
Contaminated soil	4,766	14%	735	4.9×10^{-2}	7%
Unprotected springs	3,150	9%	847	6.8×10^{-2}	8%
Tap water from piped water supply system	9	0.03%	3	1.7×10^{-4}	0.02%
Containers used to draw and store spring water and tap water	6,589	19.68%	2,266	1.9×10^{-1}	22%
Total	33,476	100%	10,172	6.8×10^{-1}	100%

[*]Effect of interventions: (A) Fencing and covering the drains reduced the disease burden from 677 to 547 per 1000 cases per year

(B) Use of greywater filters reduced the disease burden from 677 to 638 person per year

5.4   Discussion

5.4.1   Concentration of bacteria and waterborne viruses

The concentration of *E. coli*, *Salmonella* spp. and total coliforms obtained in this study were comparable to values reported for studies carried out in similar environments (Table 5.3). The factor of 0.08 used for determining the concentration of the pathogenic *E. coli* and *Salmonella* spp. obtained from the literature (Haas et al., 1999; Machdar et al., 2013) may have caused an underestimation or overestimation of the risk of infection. In future studies, identifying their isolates of *E. coli* and *Salmonella* spp. followed by serotyping and assaying for virulence factors associated with their pathogenic strains may be carried out. The detection and enumeration of *E. coli* and *Salmonella* spp. using Chromocult® Coliform Agar media was done with dilution factors of 10, 100 and 1000. However, this method does not detect non-culturable or damaged cells, even though the substrate may be specific *E. coli* or *Salmonella* spp. (Rompré et al., 2002).

The highest concentration of *E. coli* and total coliforms prevailed in grey water samples from the same location in a tertiary drain, which was an indication of the potential high public health risk from the grey water stream. The highest concentration of *Salmonella* spp. prevailed in surface water samples from the inlet of the secondary storm water drain, which may be attributed to contamination with faecal material. The detection of *E. coli* in 33% of the tap water samples could have been due to contaminated water in storage tanks (installed at households to ensure a reliable drinking water supply in case of no inflow from the distribution network) or from contaminated ground water infiltration when the pressure in the distribution network is low and with pipe leakage. Moreover, the contamination of drinking water in the distribution network and household storage in similar environments is not uncommon (Jensen et al., 2002; Oswald et al., 2007; Machdar et al., 2013).

The concentration of viruses in water and wastewater samples was used to determine the actual *E. coli* to pathogen ratio. The glass wool recovery of virus particles in the samples of 38% and a factor of 1000 to cater for up concentration of the virus particles due to the glass wool protocol (from 10 L to 10 mL) were taken into account in the determination of the virus concentration (Katukiza et al., 2013). Previous QMRA studies in urban slums used the theoretical concentration ratio of *E. coli* to virus ranging from $1:10^{-5}$ to $1:10^{-6}$ (Howard et al., 2006b; Labite et al., 2010; Machdar et al., 2013). The theoretical *E. coli* to pathogen ratio does not take into account the level of environmental pollution in slums, which may significantly underestimate or overestimate the risk of infection. Moreover, its temporal and spatial variation is large (Haas et al., 1999; Howard et al., 2006b; van Lieverloo et al., 2007).

The average concentration ratio of *E. coli* to HAdv for both grey water and surface water samples obtained in this study was comparable to the concentration ratio of *E.*

coli to viruses ranging from 2.6×10^5:1 to 2.2×10^7:1 obtained in previous studies (Lodder and de Roda Husman, 2005; Rose et al., 1996; Simpson et al., 2003; van Lieverloo et al., 2007) in different environments. In addition, the concentration ratio of *E. coli* to HAdv of 3.9×10^3:1 obtained in this study for spring water was lower than the value of 10^5:1 from the literature used by Labite et al. (2010) to estimate the disease burden associated with drinking water treatment and distribution in a slum of Accra (Ghana). The *E. coli* to virus ratio from the literature should thus only be used for global estimations of the risk of the infection and the disease burden, but cannot be used for determining the risk of infection and the disease burden for specific urban slums. This is because the survival of viruses may be enhanced by the level of contamination (Vasickova et al., 2010), which is higher in urban slum environments. In addition, the *E. coli* to pathogen ratio may change over time because *E. coli* originates from the entire population every day, whereas the viral pathogens originate from the proportion of the population that is infected at a point in time. The prevalence of viral infections varies with seasons (Armah et al., 1994). There is thus a need for further research to determine the seasonal variation of the pathogen concentration.

5.4.2 Risk of infection

The health outcome from exposure to a pathogen depends on its virulence and infectivity and on the exposure route, which is affected by the environmental system behaviour (Haas et al., 1999; Mathers et al., 2006; Pedley et al., 2006; Petterson et al., 2006). The human health effect from exposure to a hazard in the environment increases with the pathogen dose. According to the WHO (2004), the reference level of tolerable risk to human health from pathogenic enteric viruses, bacteria or protozoa is 10^{-6} DALY per person year. The maximum value of the annual risk of infection from waterborne viruses is 10^{-4} (Crabtree et al., 1997; Hamilton et al., 2006; Mena and Gerba, 2009; Venter et al., 2007).

The risk of infection in Bwaise III was constituted by waterborne viruses and bacteria in the contaminated slum to which the population was exposed. Based on the sources of contamination, this was via the water supply chain and the sanitation chain. The high annual probability of infection of 1 from exposure to HAdv and RV in surface water, grey water and contaminated spring water may be attributed to their low infection dose compared to *E. coli* O157:H7 and *Salmonella* spp. The risk of infection from exposure to the grey water ranged from 5.5×10^{-2} constituted by *Salmonella* spp. to 8.0×10^{-1} constituted by HAdv (Table 5.5). This justifies the need to prioritise grey water management in urban slums. The highest contribution to the number of infections amounted to 41% from *E. coli* O157:H7, followed by 32% from HAdv, 20% from RV and 7% from *Salmonella* spp. (Figure 5.3). The contribution to the number of infections of RV was lower than that of HAdv, although they were both detected in the two main sources of contamination (contaminated surface water and grey water) with concentrations of the same order of magnitude (Table 5.4). HAdv is considered to be more stable and prevalent than RV (Pina et al., 1998; Rigotto et al., 2011; Wyn-Jones and Sellwood, 2001) and this may have caused the difference in their relative contribution to infections.

The estimated risk of infection from different exposures varied from $2.76x10^{-6}$ to 1 (Table 5.5). The efficiency of the methods used in determining the concentration of microorganisms (viruses and bacteria) in terms of recovery and the factor for determining the concentration of *E. coli* O157:153 H7 and pathogenic strain of *Salmonella* spp. may have increased uncertainty in the estimated risk of infection. The factors that influence the magnitude of the risk of infection such as the concentration of pathogens in the environment, different exposure scenarios including dermal contact and inhalation that were not investigated in this study should be taken into account in future studies. In addition, the determination of the annual risk of infection may be improved in future studies by sampling daily and also calculating the probabilities of infection for repeated exposures by resampling daily risks of single exposures from the Monte Carlo distribution (Teunis et al., 1997). The number of cases due to infections from pathogen exposure also depended on the exposed population and the dose, which varies with time. The assumptions made in the estimation of the dose in Table 5.1 may therefore be improved through extended exposure assessment studies in similar environments. The issue of multiple episodes of infections in the exposed population may also be incorporated in risk characterisation to provide better estimates of the disease burden.

5.4.3   The disease burden in the Bwaise III slum

The total number of cases per year as a result of infections (which may or may not result in clinical illness) from reference pathogens in the sources of contamination considered in the QMRA was 33,476 (Table 5.6). This is equivalent to an annual disease burden of 677 DALYs per 1000 persons per year, which is comparable to the disease burden of 500 DALYs per 1000 persons per year obtained by Machdar et al. (2013) in the Nima slum of Accra (Ghana) and the disease burden estimates for rotavirus and *Escherichia coli* O157 in low income countries by Havelaar and Melse (2003). The disease burden for Bwaise III of $6.8x10^{-1}$ DALY per person per year was much higher than the WHO reference level of tolerable risk of $1x10^{-6}$ DALY per person per year. The annual risk from either bacteria or viruses in spring water (Table 5.4) was higher than the US EPA guideline value of one waterborne infection per 10,000 users (EPA, 1986) and therefore only tap water should be used for potable use.

The highest disease burden contribution was from exposure to surface water open drainage channels (39%) followed by that from exposure to grey water in tertiary drains (24%), storage containers (22%), unprotected springs (8%), contaminated soil (7%) and tap water (0.02%), (Table 5.6). The contribution to the disease burden by use of storage was not surprising, because a recent study by Machdar et al. (2013)

found that household storage of drinking water contributed 45% of the disease burden from poor drinking water quality in Accra (Ghana). Exclusion of other pathogens such as helminths and *Cryptosporidium* may have caused an underestimation of the disease burden in Bwaise III. In addition, there is very limited epidemiological data on urban slums and thus, the variability in the health outcomes used in the estimation of the disease burden may have caused uncertainty. Overcoming these uncertainties may result in better estimates of the disease burden that ranged from 3 to 3930 DALYs per year for different exposure routes in Bwaise III (Table 5.6).

The percentage of the total disease burden of 29% in the Katoogo zone was the highest, while the lowest percentage of the disease burden of 8% was in the Kawaala zone (Figure 5.4). Katoogo zone has the highest disease burden because the outlet of the main open drainage channel (Nsooba) with the highest pathogen concentration, the two spring water sources and most of the grey water tertiary drains in Bwaise III are located in this area (Katukiza et al., 2013). The Kawaala zone had the lowest disease burden because there are only few grey water drains in this area and the exposure to surface water in this zone is lower compared to other zones. The wide spread of the sources of contamination and the lack of intervention measures puts the entire population of Bwaise III at a high risk of infection.

5.4.4 Intervention options to reduce the risk of infection and the disease burden in Bwaise III

The proposed intervention options to reduce the disease burden in Bwaise III include those for managing exposure to hazards and others for communicating the potential risks via sanitation and water supply pathways to the public (Table 5.6). The highest disease burden was from the open storm water drainage channels and hence priority number one should be aimed at preventing contact between the population and contaminated surface water. The first option could be through providing a perimeter fence around all drains with provisional gates for operation and maintenance purposes only. The second option could be to provide lining for all storm water drains and also reinforced concrete covers that can only be removed during routine maintenance by personnel with protective clothing. Assuming a 50% efficiency of either of these options, the disease burden in Bwaise III would reduce from 677 to 547 DALYs per 1000 persons per year, equivalent to the 19% reduction in total annual number of cases for all pathogens.

Grey water accounted for the second highest contribution to the disease burden and its management needs to be prioritised if the risk of infections is to be minimised. Soil and sand filters are low cost technologies that could be implemented to reduce the concentration of water borne viruses and bacteria in grey water before discharge to the tertiary drains. This option however requires treatment at source since the housing units do not have plumbing for water supply and sewerage services. Using values for treatment efficiency of 1 log removal (90%) for viruses and 2 log removal (99%) for bacteria (Hijnen et al., 2004; Elliott et al., 2008) and assuming a 70% application of soil and sand filters at household level, the disease burden in Bwaise III reduces from 677 to 588 DALYs per 1000 persons per year (Table 5.6). This is equivalent to 11% reduction in the total annual number of cases for all pathogens and it can thus only be effective when combined with other measures to reduce the risk of infection.

As a measure to prevent the risk of infection from spring water, the local authorities should close the spring sources and encourage residents to use tap water from the existing piped water supply system, which is available at a cost. This measure however, can only reduce the disease burden by 8% (Table 5.6). Hygiene awareness campaigns as part of risk communication is an option that may be combined with any intervention to increase effectiveness to achieve the desired health outcomes. This is probably the most feasible option to minimise risk of infection from use of storage containers that accounted for 22% of the disease burden (Table 5.6).

5.5 Conclusions

This study has shown that the risk of infection from waterborne viruses and bacteria in Bwaise III slum was high. Waterborne viruses accounted for more than 50% of the mean estimated infections. The disease burden from various exposures was 10^2 to 10^5 times higher than the WHO reference level of tolerable risk of 1×10^{-6} DALYs per person per year. The major sources of microbial risks were contaminated surface water in open drainage channels, grey water in tertiary drains and storage containers contributing 39%, 24% and 22% of the total DALYs per year, respectively. The *E. coli* to pathogen ratio varied depending on the level of environmental pollution. Hence, using the literature values of this ratio to estimate the risk of infection for specific locations causes uncertainty. A combination of sanitation and hygiene interventions is required to minimise the risk of infection constituted by waterborne viruses and bacteria in the identified sources of contamination. There is a need for further research on epidemiology in urban slums to overcome the variability in health outcomes used in calculation of the disease burden. In addition, the effect of seasons in a year on viral contamination and infections requires an investigation to increase the reliability of the mean estimated risk of infection.

References

Alexander, M., Andow, D.A., Gillett, J.W., 1986. Fate and movement of microorganisms in the environment. Environmental Management 10(4), 463-493.

Alirol, E., Getaz, L., Stoll, B., Chappuis, F., Loutan, L., 2010. Urbanisation and infectious diseases in a globalised world. Lancet Infectious Diseases 10, 131-41.

APHA, AWWA, WEF., 2005. Standard Methods for the Examination of Water and Wastewater. American Public Health Association, American Water Works Association and Water Environment Federation publication, 21st edition, Washington DC, USA.

Armah, G.E, Mingle, J.A., Dodoo, A.K., Anyanful, A., Antwi, R., Commey, J., Nkrumah, F.K., 1994. Seasonality of rotavirus infection in Ghana. Annals of Tropical Paediatrics 14(3), 223-9.

Ashbolt, N.J., 2004. Microbial contamination of drinking water and disease outcomes in developing regions. Toxicology 198, 229-238.

Bezerra, P.G.M., Britto, M.C.A., Correia, J.B., Duarte, M.M.B., Fonceca, A.M., Rose, K., Hopkins, M.J., Cuevas, L.E., McNamara, P.S., 2011. Viral and Atypical Bacterial Detection in Acute Respiratory Infection in Children Under Five Years. PLoS ONE 6(4), e18928. doi:10.1371/journal.pone.0018928.

Boom, R., Sol, C., Beld, M., Weel, J., Goudsmit, J., Wertheim-van Dillen, P., 1999. Improved silica-guanidiniumthiocyanate DNA isolation procedure based on selective binding of bovine alpha-casein to silica particles. Journal of clinical microbiology 37(3), 615-619.

Boom, R., Sol, C.J., Salimans, M.M., Jansen, C.L., Wertheim-van Dillen, P.M., van der Noorda, J., 1990. Rapid and simple method for purification of nucleic acids. Journal of clinical microbiology 28, 495-503.

Butala, N.M., VanRooyen, M.J, Patel, R.B., 2010. Improved health outcomes in urban slums through infrastructure upgrading. Social Science and Medicine 71(5), 935-940.

Buttenheim, A,M., 2008. The sanitation environment in urban slums: implications for child health. Population and Environment 30, 26-47.

Byamukama, D., Kansiime, F., Mach, R.L., Farnleitner, A.H., 2000. Determination of Escherichia coli contamination with Chromocult coliform agar showed a high level of discrimination efficiency for differing fecal pollution levels in tropical waters of Kampala, Uganda. Applied and Environmental Microbiology 66, 864–868.

Carden, K., Armitage, N., Winter, K., Sichone, O., Rivett, U., Kahonde, J., 2007. The use and disposal of greywater in the non-sewered areas of South Africa: part 1-quantifying the grey water generated and assessing its quality. Water SA 33, 425-32.

Crabtree, K.D., Gerba, C.P., Rose, J.B., Haas, C.N., 1997. Waterborne Adenovirus: A risk assessment. Water Science and Technology 35, 1-6.

Cronin, A.A., Shrestha, D., Spiegel, P., Gore, F., Hering, H., 2009. Quantifying the burden of disease associated with inadequate provision of water and sanitation in selected sub-Saharan refugee camps. Journal of Water and Health 7(4), 557-568.

Dowd, S.E, Gerba, C.P., Pepper, I.L., Pillai, S.D., 2000. Bioaerosol transport modeling and risk assessment in relation to biosolid placement. Journal of Environmental Quality 29(1), 343-348.

Elliott, M.A., Stauber, C.E., Koksal, F., DiGiano, F.A., Sobsey, M.D., 2008. Reductions of *E. coli*, echovirus type 12 and bacteriophages in an intermittently operated household-scale slow sand filter. Water research 42(10-11), 2662–2670.

EPA (Environmental Protection Agency), 1986. Ambient Water Quality Criteria for Bacteria. US Environmental Protection Agency, Report No. EPA 440/5-84-002, Washington, DC.

Ferguson, C., de Roda Husman, A.M., Altavilla, N., Deere, D., Ashbolt, N., 2003. Fate and Transport of Surface Water Pathogens in Watersheds. Critical Reviews in Environmental Science and Technology 33(3), 299-361.

Fewtrell, L., Prüss-Ustün, A., Bos, R., Gore, F., Bartram, J., 2007. Water, Sanitation and Hygiene: Quantifying the Health Impact at National and Local Levels in Countries with Incomplete Water Supply and Sanitation Coverage. Environmental Burden of Disease Series No. 15. WHO, Geneva.

Haas, C.N., Rose, J.B., Gerba, C.P., 1999. Quantitative Microbial Risk Assessment, John Wiley & Sons, New York, USA.

Haas, C.N., Thayyar-Madabusia, A., Rose, J.B., Gerba, C.P., 2000. Development of a dose-response relationship for Escherichia coli O157:H7. International Journal of Food Microbiology 1748, 153–159.

Hamilton, A.J., Stagnitti, F., Premier, R., Boland, A., Hale, G., 2006. Quantitative Microbial Risk Assessment Models for Consumption of Raw Vegetables Irrigated with Reclaimed Water. Applied and Environmental Microbiology 72(5), 3284–3290.

Havelaar, A.H., Melse, J.M., 2003. Quantifying public health risk in WHO Gudelines for Drinking Water-Quality. 3720 BA Bilthoven, the Netherlands: RIVM.

Hijnen, W.A., Schijven, J.F., Bonné P., Visser, A., Medema, G.J., 2004. Elimination of viruses, bacteria and protozoan oocysts by slow sand filtration. Water Science and Technology 50(1), 147-54.

Howard, G., Jahnel, J., Frimmel, F.H., McChesney, D., Reed, B., Schijven, J., Braun-Howland E., 2006a. Human excreta and sanitation: Potential hazards and information needs. In: Schmoll, O., Howard, G., Chilton, J., Chorus, I. (Eds.): Protecting groundwater for health: managing the quality of drinking-water sources, 275-308. IWA Publishing, London, UK.

Howard, G., Pedley, S., Barret, M., Nalubega, M., Johal, K., 2003. Risk factors contributing to microbiological contamination of shallow groundwater in Kampala, Uganda. Water Research 37, 3421–9.

Howard, G., Pedley, S., Tibatemwa, S., 2006b. Quantitative microbial risk assessment to estimate health risks attributable to water supply: Can the technique be applied in developing countries with limited data? Journal of Water and Health 4(1), 49-65. http://www.microrisk.com/uploads/microrisk_qmra_methodology.pdf. [Accessed on 15th November, 2012].

Isunju, J.B., Schwartz, K., Schouten, M.A., Johnson, W.P., van Dijk, M.P., 2011. Socio-economic aspects of improved sanitation in slums: A review. Public Health 125, 368-376.

Jensen, P., Ensink, J., Jayasinghe, G., van der Hoek, W., Caincross, S., Dalsgaard, A., 2002. Domestic transmission routes of pathogens: the problem of in house contamination of drinking water during storage in developing countries. Tropical Medicine and International Health 7, 604-609.

Joyce, J., Granit, J., Frot, E., Hall, D., Haarmeyer, D., Lindström, A., 2010. The impact of the global financial crisis on financial flows to the water sector in Sub-Saharan Africa. Stockholm: SIWI.

Katukiza, A.Y., Ronteltap, M., Niwagaba, C., Kansiime, F., Lens, P.N.L., 2010. Selection of sustainable sanitation technologies for urban slums - A case of Bwaise III in Kampala, Uganda. Science of the Total Environment 409(1), 52-62.

Katukiza, A.Y., Ronteltap, M., Niwagaba, C.B., Foppen, J.W.A., Kansiime, F., Lens, P.N.L., 2012. Sustainable sanitation technology options for urban slums. Biotechnology Advances 30, 964-978.

Katukiza, A.Y., Temanu, H., Chung, J.W., Foppen, J.W.A., Lens, P.N.L., 2013. Genomic copy concentrations of selected waterborne viruses in a slum environment in Kampala, Uganda. Journal of Water and Health 11(2), 358-369.

Kemmeren, J.M., Mangen, M.J.J., Van Duynhoven. H.P.Y.T., Havelaar, A.H., 2006. Priority setting of food borne pathogens: Disease burden and costs of selection enteric pathogens. Microbiological laboratory for health protection, RIVM, The Netherlands. http://www.rivm.nl/bibliotheek/rapporten/330080001. [Accessed on 5th November, 2012].

Kulabako, N.R., Ssonko, N.K.M., Kinobe, J., 2011. Greywater Characteristics and Reuse in Tower Gardens in Peri-Urban Areas – Experiences of Kawaala, Kampala, Uganda. The Open Environmental Engineering Journal 4, 147-154.

Labite, H., Lunani, I., van der Steen, P., Vairavamoorthy, K., Drechsel, P., Lens, P., 2010. Quantitative Microbial Risk Analysis to evaluate health effects of interventions in the urban water system of Accra, Ghana. Journal of Water and Health 8(3), 417-430.

Lodder, W.J., de Roda Husman, A.M., 2005. Presence of noroviruses and other enteric viruses in sewage and surface waters in the Netherlands. Applied and Environmental Microbiology 71(3), 1453–1461.

Lopez, A.D., Mathers, C.D., Ezzati, M., Jamison, D.T., Murray, C.J.L., 2006. Global and regional burden of disease and risk factors, 2001: systematic analysis of population health data. Lancet 367, 1747-57.

Lüthi C, McConville J, Kvarnström E., 2010. Community-based approaches for addressing the urban sanitation challenges. International Journal of Urban Sustainable Development 1(1), 49-63

Machdar, E., van der Steen, N.P., Raschid-Sally, L., Lens, P.N.L., 2013. Application of Quantitative Microbial Risk Assessment to analyze the public health risk from poor drinking water quality in a low income area in Accra, Ghana. Science of the Total Environment 449, 134–142.

Mara, D., Alabaster, G., 2008. A new paradigm for low-cost urban water supplies and sanitation in developing countries. Water Policy 10(2), 119-129.

Mara, D., Lane, J., Scott, B., Trouba, D., 2010. Sanitation and Health. PLoS Medicine 7, 1-7.

Mara, D.D., Sleigh, P.A., Blumenthal, U.J., Carr, R.M., 2007. Health risks in wastewater irrigation: Comparing estimates from quantitative microbial risk analyses and epidemiological studies. Journal of Water and Health 5(1), 39-40.

Martens, W., Böhm, R., 2009. Overview of the ability of different treatment methods for liquid and solid manure to inactivate pathogens. Bioresource Technology 100(22), 5374-5378.

Mathers, C.D., Lopez, A.D., Murray, C.J.L., 2006. The burden of disease and mortality by condition: data, methods and results for 2001. In: Lopez AD, Mathers CD, Ezzati M, Jamison DT, Murray CJL, eds. Global burden of disease and risk factors, pp 45–240. New York: Oxford University Press.

Mayer, B.T., Koopman, J.S., Ionides, E.L., Pujol, J.M., Eisenberg, J.N.S., 2011. A dynamic dose–response model to account for exposure patterns in risk assessment: a case study in inhalation anthrax. Journal of the Royal Society Interface 8, 506-517.

Mena, K.D., Gerba, C.P., 2009. Waterborne adenovirus. Reviews of Environmental Contamination and Toxicology 198, 133-67.

Montgomery, M.A., Elimelech, M., 2007. Water And Sanitation in Developing Countries: Including Health in the Equation. Environmental Science and Technology 41, 17-24.

Muoki, M.A., Tumuti, D.S., Rombo, G.O., 2008. Nutrition and public hygiene among children under five years of age in Mukuru slums of Makadara Division, Nairobi. East African Medical Journal 85(8), 386-97.

Ooi, G.L., Phua, K.H., 2007. Urbanization and Slum Formation. Journal of Urban Health 84(Suppl 1), 27-34.

Oswald, W., Lescano, A., Bern, C., Calderon, M., Cabrera, L., Gilman, R., 2007. Fecal contamination of drinking water within peri-urban households, Lima Peru. American Journal of Tropical Medicine and Hygiene 77, 699-704.

Ottoson, J., Stenström, T.A., 2003. Faecal contamination of grey water and associated microbial risks. Water Research 37(3), 645-655.

Pang, X.L., Lee, B., Boroumand, N., Leblanc, B., Preiksaitis, J.K., Yu Ip. C.C., 2004. Increased detection of rotavirus using a real time reverse transcription polymerase chain reaction (RT PCR) assay in stool specimens from children with diarrhea. Journal of Medical Virology 72, 496-501.

Pedley, S., Yates, M., Schijven, J.F., West, J., Howard G, Barrett M., 2006. Pathogens: Health relevance, transport and attenuation. Protecting Groundwater for Health: Managing the Quality of Drinking-water Sources. Eds. Schmoll, O., Howard, G., Chilton, J. and Chorus, I. pp 49-76 London, UK: IWA Publishing.

Peterson, J.D., Murphy, R.R., Jin, Y., Wang, L., Nessl, M.B., Ikehata, K., 2011. Health Effects Associated with Wastewater Treatment, Reuse, and Disposal. Water Environment Research 23, 853-1875.

Petterson, S., Signor, R., Ashbolt, N., Roser, D., 2006. QMRA methodology. MicroRisk project publication. http://www.microrisk.com/uploads/microrisk_qmra_methodology.pdf. [Accessed on 15th November, 2012].

Pina, S., Puig, M., Lucena, F., Jofre, J., Girones, R., 1998. Viral pollution in the environment and in shellfish: human adenovirus detection by PCR as an index of human viruses. Applied and Environmental Microbiology 64, 3376-3382.

Rigotto, C., Hanley, K., Rochelle, P.A., De Leon, R., Barardi, C.R.M., Yates, M.V., 2011. Survival of Adenovirus Types 2 and 41 in Surface and Ground Waters Measured by a Plaque Assay. Environmental Science and Technology 45, 4145–4150.

Rodríguez, R.A., Pepper, I.L., Gerba, C.P., 2009. Application of PCR-Based Methods To Assess the Infectivity of Enteric Viruses in Environmental Samples. Applied and Environmental Microbiology 75, 297–307.

Rompré, A., Servais, P., Baudart, J., de-Roubin, M., Patrick Laurent, P., 2002. Detection and enumeration of coliforms in drinking water: current methods and emerging approaches. Journal of Microbiological Methods 49, 31–54.

Rose, J.B., Dickson, L.J., Farrah S.R., Carnahan, R.P., 1996. Removal of pathogenic and indicator microorganisms by a full-scale water reclamation facility. Water Research 30(11), 2785–2797.

Schets, F.M. Schijven, J.F., de Roda Husman, A.M., 2011. Exposure assessment for swimmers in bathing waters and swimming pools. Water Research 45(7), 2392-2400.

Schönning, C., Westrell, T., Stenström, T.A., Arnbjerg-Nielsen, K., Hasling, A.B., Høibye, L., Carlsen, A., 2007. Microbial risk assessment of local handling and use of human faeces. Journal of Water and Health 5(1), 117-128.

Simpson, D., Jacangelo, J., Loughran, P., McIlroy, C., 2003. Investigation of potential surrogate organisms and public health risk in UV irradiated secondary effluent. Water Science and Technology 47(9), 37–43.

Stanek, E.J., Calabrese, E.J., 1995. Daily estimates of soil ingestion in children. Environmental Health Perspectives 103(3), 276-285.

Stenström, T.A., Seidu, R., Ekane, N., Zurbrügg, C., 2011. Microbial Exposure and Health Assessments in Sanitation Technologies and Systems. Stockholm, Sweden: Stockholm Environment Institute, EcoSanRes Series.

Steyn, M., Jagals, P., Genthe, B., 2004. Assessment of microbial infection risks posed by ingestion of water during domestic water use and full-contact recreation in a mid-southern Africa region. Water Science and Technology 50(1), 301-308.

Templeton, M.R., Andrews, R.C., 2008. Hofmann R. Particle-Associated Viruses in Water: Impacts on Disinfection Processes, Critical Reviews in Environmental Science and Technology 38, 137-164.

Teunis, P.F.M., Medema, G.J., Kruidenier, L., Havelaar, A.H., 1997. Assessment of the risk of infection by Cryptosporidium or Giardia in drinking water from a surface water source. Water Research 31(6), 1333-1346.

UBOS, 2011. National population statistics report. Uganda National Bureau of Statistics. http://www.ubos.org. [Accessed on 12th, November 2013]

van Lieverloo, J.H.M., Mirjam Blokker, E.J., Medema, G., 2007. Quantitative microbial risk assessment of distributed drinking water using faecal indicator incidence and concentrations. Journal of Water and Health 5 (suppl.1), 131-149.

Vasickova, P., Pavlik, I., Verani, M., Carducci, A., 2010. Issues Concerning Survival of Viruses on Surfaces. Food and Environmental Virology 2, 24-34.

Venter, J.M.E., van Heerden, J., Vivier, J.C., Grabow, WOK., Taylor, M.B., 2007. Hepatitis A virus in surface water in South Africa: what are the risks? Journal of Water and Health 5, 229-239.

Victora, C.G., Smith, P.G., Vaughan, J.P., Nobre, L.C., Lombard, C., Teixeira, A.M.B., Fuchs, S.C., Moreira, L.B.B., Gigante, L.P., Barros, F.C., 1988. Water Supply, Sanitation and Housing in Relation to the Risk of Infant Mortality from Diarrhoea. International Journal of epidemiology 17(3), 651-654.

Vilaginès, Ph., Sarrette, B., Husson, G., Vilaginès, R., 1993. Glass Wool for Virus Concentration at Ambient Water pH Level. Water Science and Technology 27, 299-306.

Westrell, T., Schönning, C., Stenström, T.A., Ashbolt, N.J., 2004. QMRA (quantitative microbial risk assessment) and HACCP (hazard analysis and critical control points) for management of pathogens in wastewater and sewage sludge treatment and reuse. Water Science and Technology 50(2), 23-30.

WHO, UNICEF., 2012. Progress on drinking water and sanitation. Joint Monitoring Programme for water supply and sanitation (JMP). 1211 Geneva 27, Switzerland.

WHO., 2004. Recommendations. In Guidelines for Drinking-Water Quality, 3rd edition. Vol. 1. World Health Organization, Geneva.

Wyn-Jones, A.P., Carducci, A., Cook, N., D'Agostino, M., Divizia, M., Fleischer, J., Gantzer, C., Gawler, A., Girones, R., Höller, C., de Roda Husman, A.M., Kay, D., Kozyra, I., López-Pila, J., Muscillo, M., Nascimento, M.S.J., Papageorgiou, G., Rutjes, S., Sellwood, J., Szewzyk, R, Wyer, M., 2011. Surveillance of adenoviruses and noroviruses in European recreational waters. Water Research 45, 1025-1038.

Wyn-Jones, A.P., Sellwood, J., 2001. Enteric viruses in the aquatic environment. Journal of Applied Microbiology 91, 945-962.

Chapter 6:   Grey water characterisation and pollutant loads in an urban slum

This Chapter is based on:

Katukiza AY, Ronteltap M, Niwagaba C, Kansiime F, Lens, P.N.L., 2013. Grey water characterisation and pollutant loads in an urban slum. *International Journal of Environmental Science and Technology*. "Accepted".

Abstract

On-site sanitation provisions in urban slums rarely prioritise grey water management, yet it forms the largest fraction of wastewater. This study was carried out to characterise grey water and quantify its pollutant loads in Bwaise III (Uganda) and to provide data for grey water management in urban slums of developing countries. Samples were collected for analysis from ten representative households as well as from 4 tertiary drains and the main drainage channel for 7 months in two dry seasons. Grey water production was found to comprise 85% of the domestic water consumption. The Chemical Oxygen Demand (COD) concentration in the grey water generated by laundry, in the kitchen and in the bathroom was 9225 ± 1200 mgL^{-1}, 71250 ± 1011 mgL^{-1} and 4675 ± 750 mgL^{-1}, while the BOD$_5$ (Biochemical Oxygen Demand) to COD ratio was 0.24 ± 0.05, 0.33 ± 0.08 and 0.31 ± 0.07, respectively. The maximum concentration of *Escherichia coli* and total coliforms was 2.05×10^7 cfu.(100 mL)$^{-1}$ and 1.75×10^8 cfu.(100 mL)$^{-1}$, respectively, in grey water from the bathroom, while that of *Salmonella* spp. was 7.32×10^6 cfu.(100 mL)$^{-1}$ from laundry. Analysis of variance (ANOVA) showed a significant difference in the concentration of COD, TSS (total suspended solids), TOC (Total Organic Carbon), DOC (Dissolved Organic Carbon), TP (total phosphorus), SAR (sodium adsorption ratio), oil and grease, and *Salmonella* spp. in grey water from laundry, bathroom and kitchen ($P<0.05$). The high loads of COD (>500 kg.d^{-1}), TSS (>200 kg.d^{-1}), nutrients (8.3 kg TKN.d^{-1} and 1.4 kg TP.d^{-1}), and microorganisms (10^6 to 10^9 cfu.c^{-1}d^{-1}) originating from grey water in Bwaise III show that grey water poses a threat to the environment and a risk to human health in urban slums. Therefore, there is a need to prioritise grey water treatment in urban slums of developing countries to achieve adequate sanitation.

6.1   Introduction

The growth of slums in urban areas in developing countries causes sanitation challenges for the urban authorities. The main challenges are related to the collection as well as treatment of excreta, solid waste and wastewater for the protection of human health and the environment (Katukiza et al., 2012; Lüthi et al., 2009; Thye et al., 2011; Tilley et al., 2010). Faecal sludge management has traditionally been considered the main issue concerning improvement of sanitary conditions in urban slums because excreta are the source of many pathogens. In contrast, the grey water streams that account for the largest volumetric flux of wastewater generated in non-sewered urban slums has so far not been prioritised in sanitation provisions (Katukiza et al., 2012; Redwood, 2008). In urban slums, grey water originates mainly from laundry, bathing and kitchen activities carried out at household level. The grey water return factor (proportion of water consumption that is converted to grey water) varies from 65% to 95% (Abu Ghunmi et al., 2011, 2008; Carden et al., 2007a; Jamrah et al., 2008; Prathapar et al., 2005). The quality and quantity of grey water are influenced by the high population density, unplanned low cost housing units with limited accessibility, income level, cultural norms and type of cleaning detergents used (Eriksson et al., 2002; Kariuki et al., 2012; Morel and Diener, 2006).

Grey water discharge results in both short-term and long-term effects on the environment and human health. Soil and ground water pollution and damage to crops are caused by high concentrations of boron, sodium or surfactants, some of which may not be biodegradable (Abu-Zreig et al., 2003; Garland et al., 2000; Gross et al., 2005; Scott and Jones, 2000). In addition, nutrients in grey water may cause eutrophication whose occurrence depends on the self purification capacity of the receiving environment (Harremoës, 1998; Morel and Diener, 2006). In particular, sodium tripolyphosphate (STPP) is an ingredient of many detergents whose use has been associated with eutrophication (Köhler, 2006). Furthermore, accumulation of micro-pollutants in the environment may cause toxicity through the food chain, distort the ecological balance (Escher and Fenner, 2011; Schwarzenbach et al., 2006; Ternes and Joss, 2006) and negatively affect humans and animals after long exposure time or after bioaccumulation and biomagnification (Snyder et al., 2003). Also pathogenic microorganisms in grey water may cause diseases that result in either morbidity or mortality depending on the severity and duration of the exposure (Birks and Hills, 2007; Eriksson et al., 2002; Ottoson and Stenström, 2003). These negative effects from grey water are likely to be more severe in slums where sanitation is inadequate.

Recent studies on grey water in peri-urban settlements in developing countries focused on its characteristics in relation to its re-use potential (Carden et al., 2007a; Kariuki et al., 2012; Kulabako et al., 2011) and its management options (Armitage et al., 2009; Carden et al., 2007b). However, there is limited information on the specific pollutant loads originating from the grey water stream in typical urban slums.

Moreover, the variability of grey water quality along the tertiary drains from the generation point (households) to the open channels (surface water) under non-flooding conditions remains uninvestigated. The objectives of this study were, therefore, to characterise grey water and quantify its pollutant loads in a typical urban slum in sub-Saharan Africa, and to provide data necessary to prioritise grey water management in urban slums of developing countries.

6.2   Materials and methods

6.2.1   Study Area

This study was carried out in Bwaise III slum in Kampala city (Uganda) during the period of January to April 2010 and May to August 2012. It is a typical slum area located in a reclaimed wetland (32° 34'E and 0° 21'N) and is drained by two major open storm water channels into which tertiary drains discharge grey water and storm water (Figure 6.1). The area experiences two dry season periods from January to March and June to August. Even though there is an extensive coverage of piped water supply infrastructure in the study area, residents in the area do not necessarily access clean potable water because they cannot afford to pay. They resort to shallow ground water sources in the form of springs that are contaminated. The area is not sewered and the majority of residents use onsite sanitation in the form of elevated pit latrines (Katukiza et al., 2010). There is no wastewater management system in place.

6.2.2   Selection of households

A total of 10 households were selected after consultations with the local leaders and preliminary assessment of the current grey water practices in the study area. The following criteria were adopted: the equal representation for both tenants and landlords because of varying household size, the inclusion of households with and without children below 3 years because of contamination from bathing children and washing diapers, the willingness to cooperate and pour grey water types in separate containers, collection of samples for all grey water types during the study period, the use of only non-water borne sanitation facilities to guarantee a right estimation of grey water quantity, and the presence of a tertiary drain connected to the main drainage channel into which grey water from bathroom, kitchen and laundry activities is disposed.

6.2.3   Selection of the tertiary drains

In the study area, tertiary drains convey grey water from some households to storm water drains. Households located in places where there are no tertiary drains dispose grey water in the nearby open spaces. Two tertiary drains in the Katoogo zone and two in the St. Francis zone were monitored in this study. Figure 6.1 shows the location of the study area and the selected tertiary drains and households.

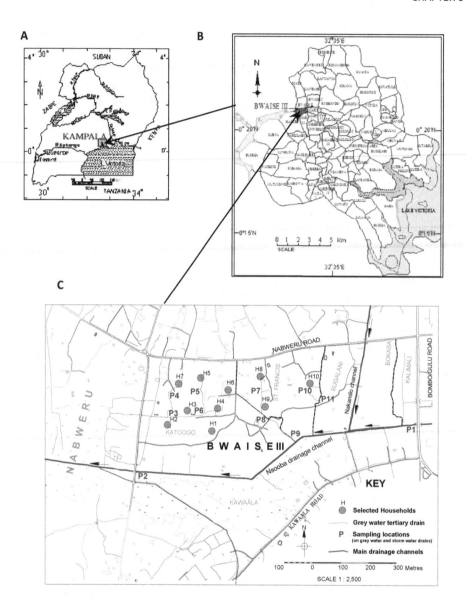

Figure 6.1: A) Map of Uganda showing the location of Kampala, B) map of Kampala showing the location of Bwaise III (Kulabako et al., 2008), and C) location of the selected households and sampling locations on grey water drains and Nsooba drainage channel in Bwaise III

6.2.4   Collection of grey water samples

Grey water samples were collected and analysed during the periods of January to April 2010 and May to August 2012. Samples for grey water from bathing, laundry and kitchen activities were collected in separate sterilised 200 mL sample bottles that were provided to households to separate their grey water types. Sampling was also done at locations P1 to P9, which are along both the tertiary drains (conveying the grey water) and the Nsooba drainage channel (Figure 6.1). Separate grey water samples were collected from households to determine the pollutant load from each source (laundry, bathroom and kitchen). In addition, sampling from tertiary drains and the main drainage channel was made in order to be able to investigate the variability of grey water quality along the tertiary drains from the generation point (households) to the open channels (surface water) under non-flooding conditions. The samples were stored in the dark at +4 °C using ice blocks and transported to the Public Health and Environmental Engineering Laboratory at Makerere University (Uganda) for analysis within one hour after sampling.

6.2.5   Analytical techniques

The pH, temperature, dissolved oxygen (DO) and electrical conductivity (EC) were measured using portable WTW (Wissenschaftlich-Technische Werkstätten) meters pH 3310; Oxi 3310 and Cond 3310. The total organic carbon (TOC) and dissolved organic carbon (DOC) of non-filtered and filtered samples, respectively, were determined using a TOC-5000A (Shimadzu, Milton Keynes, UK). Samples for measurement of DOC were filtered through 0.45µm cellulose nitrate membrane filters. To eliminate leaching of DOC from the filter itself, a control procedure was followed. The filtration apparatus was rinsed with distilled water followed by leaching of DOC from the filters using 500 mL of distilled water to ensure the DOC of the leachate was less than 0.1 mgL^{-1}. The total suspended solids (TSS) concentration was measured following standard methods (APHA, AWWA and WEF, 2005).

Chemical parameters (COD, BOD_5, TP, ortho-P, TKN, NH_4^+-N and NO_3^--N) were determined using standard methods (APHA, AWWA and WEF, 2005). COD was determined using the closed reflux colorimetric method; Kjeldahl Nitrogen was determined using the Kjeldahl method; total phosphorus was determined using persulfate digestion followed by the ascorbic acid spectrophotometric method, while ortho-phosphorus was measured using the ascorbic acid spectrophotometric method. NH_4^+-N was determined using the direct nesslerization method and NO_3^--N was determined using the calorimetric method with Nitraver 5. BOD_5 was determined using the HACH BOD track (serial no. 26197-01; Loveland, Co 80539, USA). The oil and grease concentration was obtained by solid phase extraction (SPE) using USEPA Method 1664A (USEPA, 1999). The concentration of E. coli, total coliforms and Salmonella spp. was determined with Chromocult® Coliform Agar media using the spread plate method (APHA, AWWA and WEF, 2005).

Sodium was determined using a flame photometer, while calcium and magnesium were determined using atomic absorption spectrometry (type Perkin Elmer 2380). The sodium adsorption ratio (SAR) was determined using equation 1.

$$SAR = \frac{[Na^+]}{\sqrt{([Ca^{2+}] + [Mg^{2+}])}}$$

.... (1)

Where $[Na^+]$, $[Ca^{2+}]$ and $[Mg^{2+}]$ are concentrations in mmolL^{-1} (Ganjegunte and Vance, 2006; Suarez, 1981; Paliwal and Ghandhi, 1974).

6.2.6   Pollutant loads

The specific pollutant loads from discharge of untreated grey water into the environment were determined using the average concentration of various parameters in grey water generated at household level and the grey water production per capita per day. The following expression was used:

$P_{av.c} = C_n \times Q_{av.c}$

.... (2)

Where $P_{av.c}$ is the specific pollutant load produced per capita per day for parameter n, C_n is the average concentration of parameter n in grey water and $Q_{av.c}$ is the average grey water production per capita per day, which was equal to 16.2 L.c^{-1}d^{-1} in the study area. The total pollutant loads were determined as the product of the specific pollutant loads and the contributing population.

6.2.7   Statistical analysis

IBM SPSS statistics 20 software was used for statistical analysis. One way analysis of variance (ANOVA) was used to determine the statistical (significant) differences in the concentration of various parameters in grey water from laundry, bathroom and kitchen activities at 95% confidence level. At 95% confidence interval, $p < 0.05$ means that the values of a given parameter of grey water types are significantly different. Following analysis of variances, multiple comparisons using Tukey's test were done to ascertain that values were significantly higher or significantly lower or not significantly different based on the p-values while comparing values of given parameters. The case of $p > 0.05$ means that there is no significant difference in the parameter values. (Kansiime and van Bruggen, 2001; Nsubuga et al., 2004; Mwiganga and Kansiime, 2005).

6.3   Results

6.3.1   Quantity of grey water produced in Bwaise III

The average grey water production flow in the non-sewered Bwaise III parish was 243.8 $m^3.d^{-1}$ based on the specific grey water production of 16 ± 7 $L.c^{-1}.d^{-1}$ and a population size of 15,015 obtained from the local authorities. The sources of grey water were laundry, kitchen and bathroom activities. The return factor of grey water in Bwaise III was 85 ± 7 % (Table 6.1). The average water consumption was 19.0 ± 9.7 $L.c^{-1}.d^{-1}$ while the average household size was 7 ± 3 (Table 6.1). The sources of water in Bwaise III were piped water and spring water, with more than 80% of the water demand of the selected households met by spring water because it was free compared to tap water which costed US$ 0.08 per 20 L jerry can in July, 2012. The grey water contribution from laundry, bathroom and kitchen activities was 42%, 37% and 21%, respectively.

Table 6.1: Household characteristics, water consumption and grey water production in Bwaise III

Household ID	Household characteristicts			Quantity of water consumed			Grey water produced	
	Adults	Children (< 3 years of age)	Household size	Number of 20-litre jerrycans	l/h.d	l/p.d	Grey water (l/h.d)*	Return factor**
H1	2	3	5	3	60	12	55	0.92
H2	4	4	8	6	120	15	95	0.79
H3	2	4	6	4	80	13	75	0.94
H4	7	0	7	5	100	14	85	0.85
H5	10	0	10	7	140	14	120	0.86
H6	10	4	14	10	200	14	160	0.80
H7	4	3	7	4	80	11	60	0.75
H8	2	4	6	8	160	27	120	0.75
H9	4	0	4	6	120	30	110	0.92
H10	3	0	3	6	120	40	110	0.92
Average	**5**	**2**	**7**	**6**	**118**	**19**	**99**	**0.85**
Standard deviation	**3**		**3**	**2**	**42**	**9.7**	**32**	**0.07**

* Grey water from laundry, bathroom and kitchen activities in Bwaise III accounted for 42%, 37% and 21% respectively

** Return factor = quantity of grey water produced per household per day/water consumption consumption per household per day

6.3.2 Physical and chemical characteristics of grey water in Bwaise III

The average pH of grey water from laundry, bathroom and kitchen activities in Bwaise III was 8.3±0.8, 7.6±0.4 and 7.3±1.9, respectively (Figure 6.2A). Analysis of variance (ANOVA) indicated no significant difference in the pH of grey water from laundry, bathroom and kitchen activities (p=0.127). The temperature of grey water from households and tertiary drains in Bwaise III varied from 19 °C to 29 °C, while daily ambient temperature varied from 23 °C to 29 °C. The average EC of grey water from laundry, bathroom and kitchen activities was 1671±936 µSm.cm^{-1}, 756±261 µSm.cm^{-1} and 1057±518 µSm.cm^{-1} respectively (Figure 6.2B). There was a significant difference in the EC of grey water from laundry, bathroom and kitchen activities (p=0.01). Following analysis of variance, multiple comparisons indicated that grey water from laundry had the highest EC, while the grey water from the bathroom has the lowest EC.

The average SAR of grey water from laundry, bathroom and kitchen activities in Bwaise III was 12±2, 7±3 and 8±3, respectively (Figure 6.3A). The SAR of grey water from laundry was significantly higher than the SAR of grey water generated in the bathroom and kitchen (p=0.003). The concentration of oil and grease in grey water from laundry, kitchen and bathing activities was 2.5±1.2 mg.L^{-1}, 27.5±3.7 mg.L^{-1} and 4.8±1.2 mg.L^{-1}, respectively (Figure 3B). It was significantly higher in grey water from the kitchen than in grey water from either laundry or bathroom activities (p = 0.02).

The average TOC and DOC values for grey water from laundry, kitchen and bathroom activities were 811±176 mg.L^{-1}, 677±107 mg.L^{-1}, 289±86 mg.L^{-1} and 695±171 mg.L^{-1}, 216±73 mg.L^{-1}, 533±52 mg.L^{-1} respectively (Figure 6.4A). ANOVA showed a significant difference in the concentration of TOC and DOC from laundry, bathroom and kitchen activities (p = 0.03; p = 0.03). The average concentration of TSS in grey water from laundry, bathroom and kitchen activities was 2478±1301 mgL^{-1}, 1532±633 mgL^{-1} and 3882±2988 mgL^{-1}, respectively (Figure 6.4B). The concentration of TSS in kitchen grey water was significantly higher than that in laundry and bathroom grey water (p=0.002).

Grey water from laundry activities accounted for the highest COD concentration (9225±1200 mgL^{-1}), while grey water from the bathing accounted for the lowest COD concentration (4675±170 mgL^{-1}) (Figure 6.5A). There was a significant difference in the concentration of COD from laundry, bathroom and kitchen activities in Bwaise III (p = 0.02). Grey water from laundry, bathroom and kitchen activities had a significantly higher COD than grey water from tertiary drains (p = 0.00, p=0.003, p=0.00, respectively). The average BOD$_5$ to COD ratio was 0.24±0.05, 0.31±0.07 and 0.33±0.08 for grey water originating from laundry, bathroom and kitchen activities.

Grey water from laundry activities contained the highest TP concentration (8.4±3.1 mg.L^{-1}), while grey water from the bathroom activities had the lowest TP concentration (4.3±1.9 in mg.L^{-1}) (Figure 6.5B). There was a significant difference in the concentration of TP in grey water from laundry, bathroom or kitchen activities (p = 0.003). The highest TKN concentration was 38.2 ± 5.3 mg.L^{-1} in grey water from laundry activities, while the lowest was 30.3 ± 1.5 mg.L^{-1} in grey water from kitchen activities (Figure 6.5B). There was no significant difference in the concentration of TKN in grey water from laundry, bathroom and kitchen activities in Bwaise III (p = 0.105).

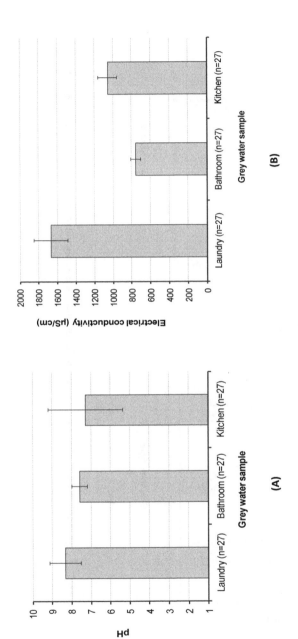

Figure 6.2:   (A) The pH and (B) electrical conductivity of grey water types generated in Bwaise III

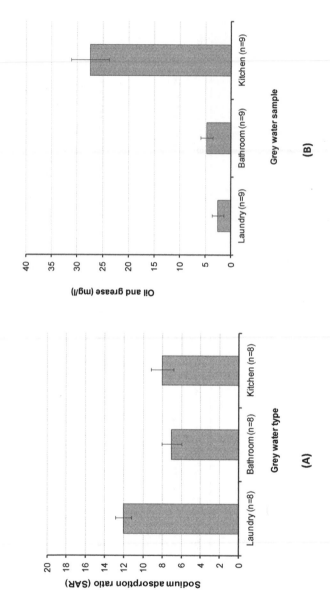

Figure 6.3: (A) Sodium adsorption ratio (SAR) and (B) oil and grease concentration in grey water from Bwaise III

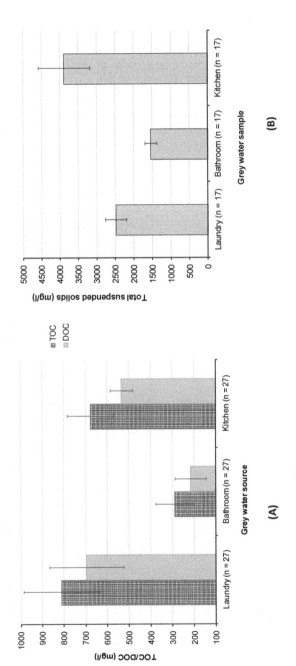

Figure 6.4: (A) Total organic carbon (TOC) and DOC and (B) total suspended solids concentration in grey water from laundry, bathing and kitchen activities in Bwaise III

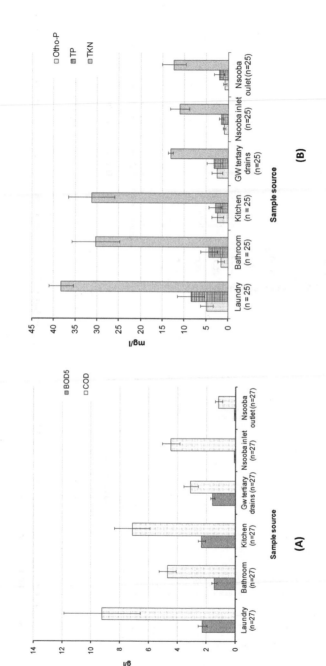

Figure 6.5: (A) Comparison of the COD and BOD₅ concentration and (B) nutrient levels of grey water from households and tertiary drains in Bwaise III, and surface water from the Nsooba channel

6.3.3   Bacteriological quality of grey water

E. coli, Salmonella spp. and total coliforms were detected, respectively, in 67%, 61% 100% of the 81 samples tested. The concentration of *E. coli* in grey water from laundry and bathroom activities in Bwaise III was 1.44×10^6 ($\pm 1.46 \times 10^6$) cfu.(100 mL)$^{-1}$ and 2.35×10^6 ($\pm 3.34 \times 10^6$) cfu.(100 mL)$^{-1}$. *E. coli* was not detected in grey water from kitchen activities (Figure 6.6). The concentration of total coliforms in grey water from laundry, kitchen and bathroom activities was 7.15×10^7 ($\pm 7.06 \times 10^7$) cfu.(100 mL)$^{-1}$, 5.87×10^7 ($\pm 8.33 \times 10^6$) cfu.(100 mL)$^{-1}$ and 1.07×10^8 ($\pm 5.87 \times 10^7$) cfu.(100 mL)$^{-1}$, respectively. The concentration of *Salmonella* spp. in grey water from laundry and bathing in Bwaise III amounted to 5.74×10^5($\pm 1.19 \times 10^6$) cfu.(100 mL)$^{-1}$ and 1.94×10^5($\pm 2.5 \times 10^5$) cfu.(100 mL)$^{-1}$, and all samples from kitchen activities tested negative for *Salmonella* spp. ANOVA showed that there was a significant difference in the *Salmonella* spp. concentration ($p = 0.03$), but no significant difference in the *E. coli* concentration in grey water from laundry and bathroom activities in Bwaise III ($p = 0.356$). In addition, there was no significant difference in the concentration of total coliforms in grey water from laundry, bathroom or kitchen activities ($p = 0.10$).

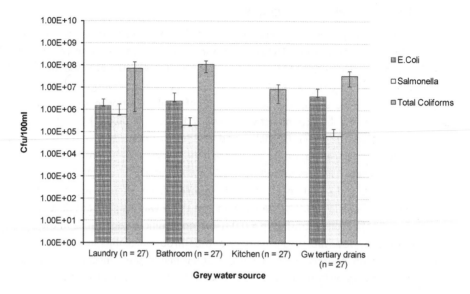

Figure 6.6:   Bacteriological quality of grey water from households and tertiary drains in Bwaise III

6.3.4 Specific pollutant loads originating from grey water

The specific pollutant loads into the environment from grey water generated at household level was determined using concentrations of the chemical and biological parameters, the average per capita water consumption per day, the return factor of 85% and the relative proportion of grey water fractions (Table 6.1). The specific pollutant loads of grey water from laundry, bathroom and kitchen activities were determined for TSS, COD, BOD_5, TP, TKN and microorganisms (Table 6.2). Grey water from laundry in Bwaise III had the highest COD load, amounting to 149 ± 37 g.c^{-1}.d^{-1}, while grey water from the kitchen had the highest TSS load amounting to 63 ± 21 g.c^{-1}.d^{-1} (Table 6.2). In addition, the highest TP and TKN loads in grey water originated from laundry activities (Table 6.2). The *E. coli* load amounted to 5.98×10^8 ($\pm 1.7 \times 10^7$) cfu.c^{-1}d^{-1} and to 9.76×10^8 ($\pm 4.5 \times 10^6$) cfu.c^{-1}d^{-1} in grey water from the bathroom and laundry, respectively, while the corresponding *Salmonella* spp. load was 4.41×10^6 ($\pm 1.3 \times 10^6$) cfu.c^{-1}d^{-1} and 7.20×10^6 ($\pm 5.9 \times 10^6$) cfu.c^{-1}d^{-1}. These values indicate that grey water from bathroom and laundry activities constitute a high public health risk in the Bwaise III slum area. The values obtained in this study are comparable to the specific grey water pollutant loads reported in the literature (Table 6.2).

Table 6.2: Comparison of Specific loads from grey water generated in Bwaise III and others parts of the world

Parameter	unit	Specific pollutant loads from grey water types in Bwaise III			Reported Pollutant loads in grey water (Busser, 2006; Friedler, 2004; Morel and Diener, 2006)*
		Laundry (n=27)	Bathroom (n=27)	Kitchen (n=27)	
TSS	$g.c^{-1}.d^{-1}$	16.8 ± 5.3	9.2 ± 3.3	13.2 ± 4.1	10 to 30
COD	$g.c^{-1}.d^{-1}$	62.6 ± 15.5	27.9 ± 7.0	24.2 ± 6.1	37 to 81
BOD_5	$g.c^{-1}.d^{-1}$	15.7 ± 5.5	13.4 ± 3.6	14.9 ± 2.9	20 to 50
TP	$g.c^{-1}.d^{-1}$	0.06 ± 0.02	0.03 ± 0.01	0.01 ± 0.002	0.2 to 6
TKN	$g.c^{-1}.d^{-1}$	0.26 ± 0.05	0.19 ± 0.07	0.10 ± 0.02	0.6 to 3.1
E.coli	$cfu.c^{-1}.d^{-1}$	$2.51 \times 10^8 (\pm 1.7 \times 10^7)$	$3.61 \times 10^8 (\pm 4.5 \times 10^7)$	-	-
Salmonella spp.	$cfu.c^{-1}.d^{-1}$	$3.02 \times 10^6 (\pm 2.9 \times 10^6)$	$1.63 \times 10^6 (\pm 1.3 \times 10^6)$	-	-
Total coliforms	$cfu.c^{-1}.d^{-1}$	$6.05 \times 10^9 (\pm 1.96 \times 10^9)$	$8.51 \times 10^9 (\pm 1.1 \times 10^9)$	$2.84 \times 10^8 (\pm 1.21 \times 10^8)$	2.4×10^9

*Compiled data based on mixed grey water studies in urban areas in USA, Asia and the middle East. No data from slums was available for comparison purposes

6.3.5   Variation of grey water quality in tertiary drains

There was variation in the quality of grey water from the source at households to the sampling locations downstream along the tertiary drains. Table 6.3 shows that the concentration of COD in mixed grey water from house households H7, H5, H8, H10 reduced by 70% to 86% in the tertiary drains, while the concentration of BOD_5 reduced by 38% to 81% in the tertiary drains and by up to 96% in the main drainage channel of Bwaise III. A similar trend was found for the total suspended solids.

The dissolved oxygen in the tertiary drains conveying grey water ranged from 0.03 mgL^{-1} to 1.87 mgL^{-1}, which was comparable to the dissolved oxygen in the main open drainage channel of less than 2 mgL^{-1} channel. The concentration of NO_3^--N varied from 1.4 mgL^{-1} to 3.5 mgL^{-1} for grey water samples from the tertiary drains and less than 0.05 mgl^{-1} in the open drainage channel. These results show that hypoxic and anoxic conditions exist in the tertiary drains, while surface water in Nsooba drainage channel is hypoxic (dissolved oxygen concentration of less than 1 mgL^{-1}).

There was no clear trend exhibited by the concentration of nutrients (TP, Ortho-P, TKN) and microorganisms (*E. coli*, *Salmonella* spp. and total coliforms) from the point of generation downstream the tertiary drains. For example, the concentration of TP increased from 6.3±1.3 mgL^{-1} to 10.2±1.7 mgL^{-1} and then decreased to 8.7±1.2 mgL^{-1} in tertiary drain 1 (Table 6.3). In tertiary drain 4, the concentration of TP increased from 5.7±0.8 mgL^{-1} to 10.5±2.1 mgL^{-1} and then decreased to 9.1±1.8 mgL^{-1}. This variation was also found for Ortho-P, NO_3^--N and TKN concentration in the monitored tertiary drains, only NH_4^+-N showed a consistent decrease downstream along the tertiary drains. In addition, the concentration of *E. coli* in tertiary drain 1 decreased from 6.3±6.3 $\log_{10}$ cfu.(100 mL)$^{-1}$ to 5.0±5.1 $\log_{10}$ cfu.(100 mL)$^{-1}$ and then increased to 6.2±6.1 $\log_{10}$ cfu.(100 mL)$^{-1}$ (Table 6.3). This variation was observed in all monitored tertiary drains for both *E. coli* and *Salmonella* spp. concentrations.

Table 6.3: Variability of grey water quality in Bwaise III along the tertiary drains from the point of generation at households to the surface water drainage channel (N= 27)*

Tertiary drain or channel	Sampling location ID	Sampling location description	COD (mg/L)	BOD$_5$ (mg/L)	TSS (mg/L)	Oil and grease (mg/L)	Sodium adsorption ratio (SAR)	NH$_4^+$-N (mg/L)	NO$_3$-N (mg/L)	TKN (mg/L)	Total-P (mg/L)	Otho-P (mg/L)	Total coliforms (log$_{10}$ cfu/100 mL)	E.coli (log$_{10}$ cfu/100 mL)	Salmonella spp. (log$_{10}$ cfu/100 mL)
Nsooba channel	P1	Inlet of Nsooba; main drainage channel	4450(±750)	95(±27)	1210(±245)	-	-	-	-	11.1(±2.2)	1.6(±0.5)	0.9(±0.3)	7.4(±7.2)	6.6(±6.4)	4.7(±4.6)
Tertiary drain no. 1	H7	Household H7 and and source of grey water	8350(±1743)	1755(±340)	3509(±1035)	35(±12)	11(±2)	25(±4)	3.0(±0.7)	38(±9)	6.3(±1.3)	3.7(±0.7)	8.0(±7.9)	6.3(±6.3)	4.5(±4.2)
	P4	Grey water tertiary drain 40 m from H7	1400(±250)	520(±45)	1100(±310)	-	-	7(±1.3)	2.5(±0.3)	12(±9)	10.2(±1.7)	5.8(±1.6)	5.8(±5.7)	5.0(±5.1)	3.4(±3.6)
	P3	Grey water tertiary drain 130 m from H7	890(±140)	385(±51)	690(±83)	-	-	8(±2.1)	1.9(±0.2)	25(±9)	8.7(±1.2)	4.7(±0.4)	7.2(±7.1)	6.2(±6.1)	4.6(±4.5)
	P2	Outlet of Nsooba; main drainage channel	1150(±145)	65(±15)	870(±119)	-	-	-	-	12.4(±2.7)	2.1(±1.2)	0.8(±0.4)	7.5(±7.6)	6.8(±6.9)	4.3(±4.1)
Tertiary drain no. 2	H5	Household H5 and and source of grey water	5800(±950)	1914(±225)	2850(±635)	22(±4)	12(±1)	40(±11)	2.8(±4)	46(±11)	6.7(±1.1)	3.7(±0.3)	7.9(±8.0)	7.0(±7.1)	4.6(±4.5)
	P5	Grey water tertiary drain 45 m from H5	1250(±167)	347.5(±46)	715(±00)	-	-	11(±1.9)	1.7(±0.3)	22(±5)	6.8(±1.4)	4.9(±0.6)	6.4(±6.3)	4.9(±4.7)	4.0(±4.2)
	P6	Grey water tertiary drain 140 m from H5	1550(±310)	578(±35)	1200(±178)	-	-	9(±1.4)	2.4(±0.7)	18(±3)	8.5(±0.9)	5(±0.7)	5.5(±5.7)	5.1(±5.1)	4.0(±3.8)
Tertiary drain no. 3	H8	Household H8 and and source of grey water	7800(±1255)	1806(±644)	3150(±835)	19(±4)	11(±3)	22(±5)	1.9(±0.4)	25(±4)	5.9(±0.4)	4.5(±0.8)	7.5(±7.3)	6.1(±6.0)	3.7(±3.6)
	P7	Grey water tertiary drain 45 m from H8	950(±145)	461(±69)	600(±75)	-	-	12(±2.1)	2.4(±1.0)	29(±7)	8.7(±1.3)	4.3(±0.5)	6.4(±6.3)	4.5(±4.4)	4.2(±4.1)
	P8	Grey water tertiary drain 140 m from H8	1050(±215)	457(±53)	620(±107)	-	-	10(±1.3)	2.7(±0.9)	17(±5)	7.3(±1.1)	5.7(±1.7)	6.1(±6.0)	5.0(±4.8)	3.2(±3.1)
	P9	Sampling point 280 m from H8 at junction of the grey water tertiary drain and Nsooba channel	1350(±125)	342(±47)	967(±315)	-	-	-	-	-	2.9(±0.6)	1.3(±0.7)	7.8(±7.9)	5.7(±5.8)	3.8(±3.7)
Tertiary drain no. 4	H10	Household H10 and and source of grey water	4300(±1760)	903(±71)	1801(±765)	28(±7)	10(±1)	31(±8)	3.1(±1.0)	30(±9)	5.7(±0.8)	3.5(±1.1)	7.8(±7.7)	4.9(±4.9)	3.7(±3.6)
	P10	Grey water tertiary drain 45 m from H10	1300(±275)	598(±62)	849(±161)	-	-	8.7(±1.6)	2.1(±0.3)	21(±6)	10.5(±2.1)	5.2(±0.9)	6.1(±6.0)	5.0(±4.9)	4.2(±4.2)
	P11	Grey water tertiary drain 140 m from H10	1250(±147)	563(±44)	775(±207)	-	-	7.6(±1.3)	2.4(±0.5)	15(±3)	9.1(±1.8)	4.9(±0.4)	5.7(±5.6)	5.2(±5.1)	4.8(±4.6)

*Grey water from households is mixed and originates from Laundry, bathroom and Kitchen activities. In brackets is the standard deviation.

6.4   Discussion

6.4.1   Grey water production in Bwaise III

The amount of grey water generated in Bwaise III was equivalent to 85±7% of the water consumption (Table 6.1). This represents a significant volume of waste water discharged into the environment. Although the specific water consumption of 19±9.7 $L.c^{-1}d^{-1}$ is below the minimum recommended global value of 30 $L.c^{-1}d^{-1}$ (WHO, 2012) and the minimum national unit water consumption of 20 $L.c^{-1}d^{-1}$, the high population density and high return factor leads to high volumetric fluxes of grey water discharge. The grey water return factor and water consumption in Bwaise III were within the reported range for peri-urban areas in sub-Saharan Africa of 75 to 95% and 15 to 30 litres per capita per day, respectively (Alderlieste and Langeveld, 2005; Carden et al., 2007a; Morel and Diener, 2006).

Laundry and bathroom activities contributed the highest grey water volume generated in Bwaise III (Table 6.1). Kulabako et al. (2011) also found, based on household interviews that the bulk of grey water in peri-urban Kampala was from laundry activities. In contrast, grey water studies in peri-urban areas of other parts of the world showed that 55% to 60% of the grey water originated from either bathroom or showering (Busser, 2006; Jamrah et al., 2008), while others found that wash basins and laundry machines each contribute about 16% to 33% (Christova-Boal et al., 1996; Friedler et al., 2004). The volume of grey water produced is thus dependent on the standard of living and sanitary infrastructure.

6.4.2   Biodegradability of grey water

The BOD_5 to COD ratio of grey water from selected households in Bwaise III ranging from 0.24 to 0.34 and the BOD_5 to TOC ratio of 1.2±0.7 for mixed grey water show that it is not easily biodegradable. A BOD_5 to COD ratio of a wastewater stream close to or above 0.5 is an indication of its good biodegradability (Hernández et al., 2007; Li et al., 2009; Metcalf and Eddy, 2003). The typical values of BOD_5 to COD ratio of untreated municipal wastewater range from 0.3 to 0.8 and the corresponding BOD_5 to TOC ratio ranges from 1.2 to 2.0 (Crites and Tchobanoglous, 1998; Metcalf and Eddy, 2003). The BOD_5 to COD ratio values obtained in this study were within the reported range of 0.21 to 0.35 on grey water (Al-Jayyousi, 2003; Eriksson et al., 2002; Jefferson et al., 2004).

The COD:N:P ratios of grey water from households (Table 6.4) indicate that grey water is deficient in nitrogen and phosphorus when compared with a value of 100:20:1 required for aerobic biological treatment (Metcalf and Eddy, 2003). These data on grey water types from Bwaise III are important in the selection of grey water treatment technologies for slums. The COD:N:P ratios were comparable to values obtained for grey water from urban areas in other parts of the world (Table 6.4). This study did not include an inventory of household chemicals used in Bwaise III, which may further detail the poor biodegradability of grey water.

Table 6.4: Comparison of the COD:N:P ratio of grey water from Bwaise III and urban areas in other regions

Grey water source	COD:N:P	Reference
Laundry	100 : 0.63 : 0.09	This study
Bathroom	100 : 1.09 : 0.09	
Kitchen	100 : 0.56 : 0.04	
Laundry, bathroom and Kitchen	100 : 1.9 : 0.07	
Shower	100 : 2.91 : 0.05	Jefferson et al., 2004
Bathroom	100 : 2.25 : 0.06	
Handwash basin	100 : 1.77 : 0.06	
Shower, bathroon, handwash basin	100 : 1.94 : 0.08	
Laundry, bathroom and Kitchen	100 : 1.65 : 1.28	Palmquist and Hanaeus, 2005
Laundry, bathroom and Kitchen	100 : 1.82 : 0.76	Krishnan et al., 2008

6.4.3 Variation of the grey water quality from tertiary drains in Bwaise III

The concentration of COD, TSS and nutrients varied along the tertiary drains. There was a 70% to 86% reduction in concentration of COD and TSS in grey water along the grey water tertiary drains from the generation sources at households to open channels (surface water). This was attributed to the sedimentation in the drains under dry weather flow conditions. Moreover, 61 (±6) % of the measured total COD concentration was in particulate form (Figure 6.5A). The concentration of nutrients varied along the tertiary drains without a clear trend (Table 6.3). This was probably due to the presence of other households downstream that discharge grey water in the same tertiary drains. Moreover, the concentration of microorganisms upstream and downstream along the tertiary drains was high (Table 6.3), which makes the stagnation of grey water in the tertiary drain in Bwaise III a major public health concern. Exposure to pathogens at points of grey water stagnation along the tertiary drains may be via involuntary ingestion of grey water especially by the children. It can also be through consumption of contaminated food since it is common in Bwaise III for families to prepare food near the drains because of limited space.

6.4.4   Grey water pollutant loads and its potential environmental impacts

The high COD load (> 500 kg.d^{-1}) of grey water generated in Bwaise III may induce hypoxic conditions in the main open surface water drainage channel. Grey water from laundry activities had the highest specific COD load (Table 6.2) because of re-using the same water for many cycles. This makes the contribution of laundry activities to the strength of grey water in Bwaise III considerable. In addition, the COD concentration of grey water generated in Bwaise III (Figure 6.5) was higher than typical values ranging from 250 mgL^{-1} to 1000 mgL^{-1} for untreated sewage (Metcalf and Eddy, 2003). Moreover, all the grey water types had a higher COD concentration (> 4000 mgL^{-1}) than the discharge standard of 100 mgL^{-1} set by the National Environmental Management Authority (NEMA) in Uganda. Grey water generated in places such as slums with low water consumption (< 30 l.c^{-1}.d^{-1}) is more concentrated than grey water from affluent or higher income areas (Morel and Diener, 2006). The low water consumption in Bwaise III thus contributes to the high COD concentration in grey water, while the high grey water return factor increased the COD load (Table 6.1).

The average nutrient loads into the environment from grey water generated in Bwaise III were 8.2 kg.day^{-1} of TKN and 1.4 kg.day^{-1} of TP (Table 6.2). Macro-nutrients (nitrogen and phosphorous) are relevant for biological treatment processes and important for plant growth. However, they have a potential negative impact on the aquatic environment. The large quantities of grey water generated in Bwaise III contribute to significant loads of phosphorous and nitrogen that may cause eutrophication in the receiving water bodies downstream of the slum (Nyenje et al., 2010). In addition, nutrient fluxes may also contaminate ground water through leaching in Bwaise III, where the water table is high (less than 1 m).

The concentration of oil and grease in grey water from Bwaise III is a major concern. Grey water especially from the kitchen had an oil and grease concentration exceeding the maximum discharge standard for wastewater effluents of 10 mgL^{-1} in Uganda (Figure 6.3B). Therefore, current practice of grey water disposal in the open in Bwaise III without treatment may reduce soil aeration and also cause infiltration problems in Bwaise III (Mohawesh et al., 2013). Changes in the chemical composition of soil have been reported as a result of irrigation with effluents rich in petrochemical products (Sharma et al., 2012). Travis et al. (2008) found that oil and grease cause soil water repellency and also form translucent films in the surface water, which interfere with aquatic life. These effects may occur in the receiving environment to which the Nsooba surface water channel discharges downstream of Bwaise III. The concentration of oil and grease obtained in this study was lower than the values reported in the literature (Carden et al., 2007; Christova-Boal et al., 1996; Travis et al., 2008). This is attributed to the differences in kitchen activities and the living standards for affluent areas and also developed countries compared to urban slums. Concentrations of oil and grease ranging from 8 to 78 mgL^{-1} were observed in laundry grey water by Christova-Boal et al. (1996). The concentration of oil and grease in the grey water thus depends on household activities and eating habits.

The infiltration or re-use for irrigation of grey water from Bwaise III may result in soil salinisation. The detrimental effects of sodium salinity and sodicity are determined based on the proportion of sodium (Na^+), calcium (Ca^{2+}) and magnesium (Mg^{2+}) in grey water or the values of SAR (Kariuki et al., 2012). The SAR values ranging from 4 to 15 obtained in this study are comparable to SAR values obtained for grey water in previous studies from peri-urban areas ranging from 2 to 12 (Kariuki et al., 2012; Kulabako et al., 2011). SAR values exceeding 50 may be obtained depending on the sodium salts used to produce detergents used at households (Friedler, 2004, Patterson, 2001). High SAR values (>6) reduce soil aeration and permeability (Morel and Diener, 2006). In addition, an increase in SAR as a result of long-term land application of grey water negatively affects soil properties and cause die off of some plant species (Gross et al., 2005). The high SAR (>6) is an indication that disposal of untreated grey water may cause long-term effects of salinity, low water infiltration rates and water clogging in the Bwaise III slum if grey water management is prioritised.

6.4.5 Microorganism concentration and loads in grey water

The high concentrations of *E. coli* and *Salmonella* spp. in grey water from households and tertiary drains in Bwaise III pose a health risk to the slum dwellers (Figure 6.6). The 10^4 to 10^8 cfu. $(100 \text{ mL})^{-1}$ concentration of microorganisms constitutes a health risk because about 8% of the determined concentrations of *E. coli* and *Salmonella* spp. are pathogenic (Haas et al., 1999; Westrell, 2004). Grey water generated in Bwaise III is therefore unfit for direct non-potable reuse and its stagnation in the tertiary drains that have low slopes creates a potential source of exposure to water borne pathogens. Despite the studies that have been carried out in peri-urban areas of sub-Saharan Africa (Armitage et al., 2009; Carden et al., 2007a; Kariuki et al., 2012; Kulabako et al., 2011), the data on microorganism loads in grey water from slums is very limited. The specific loads of microorganisms in grey water from laundry, bathroom and kitchen activities in Bwaise III were determined in this study (Table 6.2). Microorganism loads are useful in assessing the health effects during epidemiological studies and for controlling the source of pathogens to which slum dwellers are exposed.

The use of indicator organisms to determine the bacteriological quality of grey has limitations. Grey water in tertiary drains may be contaminated with many types of pathogens from different point and diffuse sources in the slum areas. The absence of indicator organisms in grey water samples may therefore not reflect the distribution of pathogens in grey water. For example, waterborne viruses were detected in grey water samples that tested negative for *E. coli* from locations along tertiary drains in the Bwaise III slum (Katukiza et al., 2013). There is thus a need for the use of other pathogen detection techniques like quantitative polymerase chain reaction (qPCR). In addition, there is a need for a Quantitative Microbial Risk Assessment (QMRA) to determine the magnitude of the microbial risks to human health from exposure to grey water. Such a QMRA approach would also support the local authorities in making decisions on the measures to reduce the microbial risk and the disease burden in urban slums.

6.5 Conclusions

Grey water from laundry, kitchen and bathroom activities in Bwaise III accounted for 85% of the domestic water consumption. The high concentration of disease causing microorganisms ranging from 10^4 to 10^8 cfu.$(100 \text{ mL})^{-1}$ poses a health risk and makes grey water not suitable for direct non-potable reuse although it contains nutrients. The high COD (> 4000 mgL^{-1}) and poor bacteriological quality makes grey management an essential provision in slum sanitation. The low BOD$_5$ to COD ratio of 0.24 to 0.34 indicates a low biodegradability of grey water. Pre-treatment of grey water is essential to avoid clogging of the main treatment unit because of high TSS concentrations (>1500 mgL^{-1}) and the presence of oil and grease. The high organic load, accumulation of sodium, oil and grease, and microorganism loads from disposal of untreated grey water in Bwaise III poses a potential environmental and health risk. Therefore, there is need for research to quantify the impacts on the environment especially the ecosystem downstream urban slums and the public health risks from the grey water pollutant loads generated in urban slums in developing countries.

References

Abu Ghunmi, L., Zeeman, G., van Lier, J., Fayyed, M., 2008. Quantitative and qualitative characteristics of grey water for reuse requirements and treatment alternatives: the case of Jordan. Water Science and Technology 58(7), 1385-1396.

Abu Ghunmi, L.A., Zeeman, G., Fayyad, M., Van Lier, J.B., 2011. Grey Water Treatment Systems: A Review. Critical Reviews in Environmental Science and Technology 41, 657-698.

Abu-Zreig, M., Rudra, R.P., Dickinson, W.T., 2003. Effect of application of surfactants on hydraulic properties of soils. Biosystems Engineering 84, 363-372.

Alderlieste, M.C., Langeveld, J.G., 2005. Wastewater planning in Djenné, Mali. A pilot project for the local infiltration of domestic wastewater. Water Science and Technology 51(2), 57-64.

Al-Jayyousi, O., 2002. Focused environmental assessment of greywater reuse in Jordan. Environmental Engineering and Policy 3(1), 67-73.

Al-Jayyousi, O.R., 2003. Greywater reuse: towards sustainable water management. Desalination 156(1), 181-192.

APHA, AWWA and WEF, 2005. Standard Methods for the Examination of Water and Wastewater. *American Public Health Association, American Water Works Association and Water Environment Federation publication,* 21st edition. Washington DC, USA.

Armitage, N.P., Winter, K., Spiegel, A., Kruger, E., 2009. Community-focused greywater management in two informal settlements in South Africa. Water Science and Technology 59(12), 2341-2350.

Birks, R., Hills, S., 2007. Characterisation of Indicator Organisms and Pathogens   in Domestic Greywater for Recycling. Environmental Monitoring and Assessment, 129, 61-69.

Busser, S., 2006. Studies on domestic wastewater flows in urban and peri-urban Hanoi. http://ir.library.osaka.ac.jp/dspace/bitstream/11094/13204/1/arfyjsps2006_395.pdf [Accessed on 7th February, 2013].

Carden, K., Armitage, N., Sichone O., Winter, K., 2007b. The use and disposal of grey water in the non-sewered areas of South Africa: Part 2- Grey water management options. Water SA, 33 (4), 433-442.

Carden, K., Armitage, N., Winter, K., Sichone, O., Rivett, U., Kahonde, J., 2007a. The use and disposal of grey water in the non-sewered areas of South Africa: Part 1- Quantifying the grey water generated and assessing its quality. Water SA 33(4), 425-432.

Chaggu, E.J., Mashauri, A., Van Buren, J., Sanders, W., Lettinga, J., 2002. PROFILE: Excreta Disposal in Dar es Salaam. Environment Management 30(5), 609-620.

Christova-Boal, D., Eden, R.E., McFarlane, S., 1996. An investigation into greywater reuse for urban residential properties. Desalination 106, 391-397.

Crites, R.W., Tchobanoglous, G., 1998. Small and Decentralized Wastewater Management Systems. McGraw-Hill Companies, California, ISBN 0-07-289087-8 USA.

Eriksson, E., Auffarth, K., Henze, M., Ledin, A., 2002. Characteristics of grey wastewater. Urban Water 4(1), 85-104.

Escher, B.I., and Fenner, K., 2011. Recent Advances in Environmental Risk Assessment of Transformation Products. Environmental Science and Technology 45, 3835-3847.

Friedler, E., 2004. Quality of individual domestic greywater streams and its implication for onsite treatment and reuse possibilities. Environmental Technology 25(9), 997-1008.

Friedler, E., Kovalio, R., Galil, N.I., 2005. On-site greywater treatment and reuse in multi-storey buildings. Water Science and Technology 51(10), 187-194.

Ganjegunte, G.K.,Vance, G.F., 2006. Deviations from the empirical Sodium adsorption ratio (SAR) and exchangeable sodium percentage (ESP) relationship. Soil Science 171(5), 364-373.

Garland, J.L., Levine, L.H., Yorio, N.C., Adams, J.L. and Cook, K.L., 2000. Graywater processing in recirculating hydroponic systems: Phytotoxicity, surfactant degradation, and bacterial dynamics.Water Research 34, 3075-3086.

Gross, A., Azulai, N., Oron, G., Ronen, Z., Arnold, M., Nejidat, A., 2005. Environmental impact and health risks associated with greywater irrigation: a case study. Water Science & Technology 52 (8), 161-169.

Haas, C.N., Rose, J.B., Gerba, C.P., 1999. Quantitative Microbial Risk Assessment, John Wiley & Sons, New York, USA.

Harremoës P., 1998. Stochastic Models for Estimation of Extreme Pollution from Urban Runoff. Water Research, 22(8), 1017-1026.

Henriques, J.J, Louis, G.E., 2011. A decision model for selecting sustainable drinking water supply and greywater reuse systems for developing communities with a case study in Cimahi, Indonesia. Journal of Environmental Management 92, 214-222.

Hernández, L, Zeeman, G., Temmink, H., Buisman, C., 2007. Characterization and biological treatment of greywater. Water Science and Technology 56(5), 193–200.

Jamrah, A., Al-Futaisi, A., Prathapar, S., Harrasi, A.A., 2008. Evaluating greywater reuse potential for sustainable water resources management in Oman. Environmental Monitoring and Assessment 137(1-3), 315-327

Jefferson, B., Burgess, J.E., Pichon, A., Harkness, J., Judd, S.J., 2001. Nutrient addition to enhance biological treatment of greywater. Water Research 35 (11), 2702-2710

Jefferson, B., Palmer, A., Jeffrey, P., Stuetz, R., Judd, S., 2004. Grey water characterisation and its impact on the selection and operation of technologies for urban reuse. Water Science and Technology 50(2), 157-164.

Kansiime, F., van Bruggen, J.J.A., 2001. Distribution and retention of faecal coliforms in the Nakivubo wetland in Kampala, Uganda. Water Science Technology 44(11-12), 19-26.

Kariuki, F.W., Ng`ang`a, V.G., Kotut, K., 2012. Hydrochemical Characteristics, Plant Nutrients and Metals in Household Greywater and Soils in Homa Bay Town. The Open Environmental Engineering Journal 5, 103-109.

Katukiza, A.Y. Ronteltap, M., Niwagaba, C., Kansiime, F., Lens, P.N.L., 2010. Selection of sustainable sanitation technologies for urban slums - A case of Bwaise III in Kampala, Uganda, Science of the Total Environment 409(1), 52-62.

Katukiza, A.Y., Ronteltap, M., Niwagaba, C.B., Foppen, J.W.A. Kansiime, F., Lens, P.N.L., 2012. Sustainable sanitation technology options for urban slums. Biotechnology Advances 30, 964-978.

Katukiza, A.Y., Temanu, H., Chung, J.W., Foppen, J.W.A., Lens, P.N.L., 2013. Genomic copy concentrations of selected waterborne viruses in a slum environment in Kampala, Uganda. Journal of Water and Health 11(2), 358-369.

Köhler, J., 2006 Detergent Phosphates: An EU Policy Assessment. Journal of Business Chemistry 3(2), 15-30.

Krishnan, V., Ahmad, D., Jeru, J.B., 2008. Influence of COD:N:P ratio on dark greywater treatment using a sequencing batch reactor. Journal of Chemical Technology and Biotechnology 83, 756-762.

Kulabako, N. R., Ssonko, N.K.M, Kinobe, J., 2011. Greywater Characteristics and Reuse in Tower Gardens in Peri-Urban Areas – Experiences of Kawaala, Kampala, Uganda. The Open Environmental Engineering Journal 4, 147-154.

Kulabako, N.R., Nalubega, M., Thunvik, R., 2008. Phosphorus transport in shallow groundwater in peri-urban Kampala, Uganda: results from field and laboratory measurements. Environmental Geology 53, 1535–1551.

Li, F., Wichmann, K, Otterpohl, R., 2009. Review of technological approaches for grey water treatment and reuses. Science of the Total Environment 407(11), 3439–49.

Lüthi, C., McConville, J., Kvarnström, E., 2009. Community-based approaches for addressing the urban sanitation challenges. International Journal of Urban Sustainable Development 1(1), 49-63.

Metcalf, Eddy., 2003. Wastewater engineering: treatment and reuse. 4th ed. 1221 Avenue of Americas, New York, NY 10020, USA: McGraw-Hill Companies.

Mohawesh, O., Mahmoud, M., Janssen, M., Lennartz, B., 2013. Effect of irrigation with olive mill wastewater on soil hydraulic and solute transport properties. International Journal of Environmental Science and Technology, DOI: 10.1007/s13762-013-0285-1.

Morel, A., Diener, S., 2006. Greywater Management in Low and Middle-Income Countries. Review of different treatment systems for households or neighbourhoods. http://www.eawag.ch/forschung/sandec/publikationen/ewm/dl/GW_managem ent.pdf [Accessed on January 3rd, 2013].

Mwiganga, M., Kansiime, F., 2005. The impact of Mpererwe landfill in Kampala–Uganda, on the surrounding environment. Physics and Chemistry of the Earth 30, 744–750.

Nsubuga, F.B., Kansiime, F., Okot-Okumu, J., 2004. Pollution of protected springs in relation to high and low density settlements in Kampala—Uganda. Physics and Chemistry of the Earth 29, 1153–1159.

Nyenje, P.M., Foppen, J.W., Uhlenbrook, S., Kulabako, R., Muwanga, A., 2010. Eutrophication and nutrient release in urban areas of sub-Saharan Africa: a review. Science of the Total Environment 408(3), 447-455.

Ottoson, J., Stenström, T.A., 2003. Faecal contamination of grey water and associated microbial risks. Water Research 37(3), 645-655.

Paliwal, K.V., Gandhi, A.P., 1974. Effect of salinity, SAR, Ca:Mg ratio in irrigation water, and soil texture on the predictability of exchangeable Sodium percentage. Soil Science 122(2), 85-90.

Palmquist, H., Hanaeus., 2005. Hazardous substances in separately collected grey- and blackwater from ordinary Swedish households. Science of the Total Environment 348, 151-163.

Paterson, C., Mara, D., Cutis, T., 2007. Pro-poor sanitation technologies. Geoforum 38(5), 901-907.

Patterson, R.A., 2001. Wastewater quality relationships with reuse options. Water Science and Technology 43 (10), 147–154.

Prathapar, S.A., Jamrah, A., Ahmed, M., Al Adawi, S., Al Sidairi, S., Al Harassi, A., 2005. Overcoming constraints in treated greywater reuse in Oman. Desalination 186(1-3), 177-186.

Redwood, M., 2008. The application of pilot research on greywater in the Middle East North Africa region (MENA). International Journal of Environmental Studies 65(1), 109-117.

Schwarzenbach, R.P., Escher, B.I., Fenner, K., Hofstetter, T.B., Johnson, C.A., von Gunten, U., Wehrli, B., 2006. The Challenge of Micropollutants in Aquatic Systems. Science 313, 1072-1077.

Scott, M. J., Jones, M. N., 2000. The biodegradation of surfactants in the environment. Biochim. Biophys. Acta, Biomembr. 1508 (1-2), 235-251.

Sharma, N.K., Bhardwaj, S., Srivastava, P.K., Thanki, Y.J., Gadhia, P.K., Gadhia, M., 2012. Soil chemical changes resulting from irrigating with petrochemical effluents. International Journal of Environmental Science and Technology 9(2), 361-370.

Snyder, S.A., Westerhoff, P., Yoon, Y., and Sedlak, D.L., 2003. Pharmaceuticals, Personal Care Products, and Endocrine Disruptors in Water: Implications for the Water Industry. Environmental Engineering Science 20(5), 449-469.

Suarez, D.L., 1981. Relation Between pH and Sodium Adsorption Ratio (SAR) and an Alternative Method of Estimating SAR of Soil or Drainage Waters. Soil Science Society of America Journal, 45(3), 469-475.

Ternes, T., Joss, A., 2006. Human pharmaceuticals, hormones andfragrances: the challenge of micropollutants in urban water management. IWA, London.

Thye, Y.P., Templeton, M.R., Ali, M., 2011. A Critical Review of Technologies for Pit Latrine Emptying in Developing Countries, Critical Reviews in Environmental Science and Technology 41(20), 1793-1819.

Tilley, E., Zurbrüg, C., Lüthi, C., 2010. A flowstream approach for sustainable sanitation systems. In: van Vliet, B., Spaargaren, G., Oosterveer, P. (eds.). Social perspectives on the Sanitation Challenge. Springer, Dordrecht, pp. 69–86.

Travis, M.J., Weisbrod, N., Gross, A., 2008. Accumulation of oil and grease in soils irrigated with greywater and their potential role in soil water repellency. Science of the Total Environment 394, 68-74.

USEPA, 2009. Method 1664 revision A: N-hexane extractable material (HEM; oil and grease) and silica gel treated N-hexane extractable material (SGT-HEM; non-polar material) by extraction and gravimetry. Washington DC: United States Environmental Protection Agency.

Westrell T., Schönning C., Stenström T.A., Ashbolt N.J., 2004. QMRA (quantitative microbial risk assessment) and HACCP (hazard analysis and critical control points) for management of pathogens in wastewater and sewage sludge treatment and reuse. Water Science and Technology 50(2), 23-30.

WHO, UNICEF, 2012 Progress on drinking water and sanitation. Joint Monitoring Programme for water supply and sanitation (JMP). 1211 Geneva 27, Switzerland

Chapter 7:   Grey water treatment in urban slums by a filtration system: optimisation of the filtration medium

This Chapter is based on:

Katukiza, A.Y., Ronteltap, M., Niwagaba ,C., Kansiime, F., Lens, P.N.L., 2013. Grey water treatment in urban slums by a filtration system: optimisation of the filtration medium. *Journal of Environmental Management*. "Under review".

Abstract

Two uPVC columns (outer diameter 160 cm, internal diameter 14.6 cm and length 100 cm) were operated in parallel and in series to simulate grey water treatment by media based filtration at unsaturated conditions and constant hydraulic loading rates (HLR). Grey water from bathroom, laundry and kitchen activities was collected from 10 households in the Bwaise III slum in Kampala (Uganda) in separate containers, mixed in equal proportions followed by settling, prior to transferring the influent to the tanks. Column 1 was packed with lava rock to a depth of 60 cm, while column 2 was packed with lava rock (bottom 30 cm) and silica sand, which was later replaced by granular activated carbon (top 30 cm) to further investigate nutrient removal from grey water. Operating the two filter columns in series at a HLR of 20 cm/day resulted in a better effluent quality than at a higher (40 cm/day) HLR. The COD removal efficiencies by filter columns 1 and 2 in series amounted to 90% and 84% at HLR of 20 cm/day and 40 cm/day, respectively. TOC and DOC removal efficiency amounted to 77% and 71% at a HLR of 20 cm/day, but decreased to 72% and 67% at a HLR of 40 cm/day, respectively. The highest log removal of *E. coli*, *Salmonella* spp. and total coli forms amounted to 3.68, 3.50 and 3.95 at a HLR of 20 cm/day respectively. The overall removal of pollutants increased with infiltration depth, with the highest pollutant removal efficiency occurring in the top 15 cm layer. Grey water pre-treatment followed by double filtration using coarse and fine media has the potential to reduce the grey water pollution load in slum areas by more than 60%.

7.1   Introduction

The uncontrolled disposal of untreated grey water generated by households in urban slums is a major sanitation issue. Grey water from urban slums contains high pathogen loads (Carden et al., 2007; Katukiza et al., 2013a; Rodda et al., 2011) and can be categorised as high strength wastewater because of its high COD concentration (> 2 g/l) (Kulabako et al., 2011; Sall and Takahashi, 2006). In addition, grey water generally contains macro-nutrients (Eriksson et al., 2002; Jefferson et al., 2001), organic micro-pollutants from household chemicals, as well as pharmaceuticals and hormones (Alder et al., 2006; Hernández-Leal et al., 2011). Laundry and personal care products used at household level and ingredients of detergents like surfactants may also be present in grey water (Garland et al., 2000; Temmink and Klapwijk, 2004; Ying, 2006). The grey water pollution load in urban slums is high, despite the low water consumption because of high pollutant concentrations due to multiple water use within the slum households.

In arid and water stressed areas, grey water is treated for reuse to reduce the demand on piped water supply systems and also to promote the use of high quality water for potable uses only (Al-Hamaiedeh and Bino, 2010; Mandal et al., 2011). In contrast, the priority in urban slums is to reduce the grey water pollution load that poses public health risks and a threat to the environment (Katukiza et al., 2013b). The negative environmental impacts from grey water discharge in urban slums may include reduced soil hydraulic conductivity, grey water ponding, die-off of some plants and hypoxic conditions in surface water drains (Gross et al., 2005; Morel and Diener, 2006; Travis et al., 2008). Moreover, the current practice of discharging untreated grey water into nearby open spaces or adjacent storm water drains in urban slums is unhygienic. It creates exposure routes to pathogens such as rotavirus and *E. coli* that cause diseases like gastroenteritis and diarrhoea, especially in the immuno-compromised groups like children under the age of 5 years and the elderly. Indeed, grey water in urban slums is contaminated with rotavirus, adenoviruses, *E. coli* and helminth eggs (Katukiza et al., 2013a; Labite et al., 2010; Sall and Takahashi, 2006).

Although various grey water treatment systems have been implemented in developing countries (Mandal et al., 2011; Morel and Diener, 2006; Rodda et al., 2011), there has been hardly any intervention at household level in non-sewered urban slums. The aim of carrying out filter column experiments in the present study was, therefore, to provide a basis for treatment of household grey water with a suitable medium (sand, crushed lava rock) filter system in an urban slum setting. Recent column studies carried out with artificial or low strength grey water (Dalahmeh et al., 2012; López-Zavala, 2007) investigated bark, sand, mulch and gravel as filtration media. In other studies, mulch and grey water towers were used to investigate the treatment of high strength grey water from peri-urban and urban areas (Kulabako et al., 2011; Tandlich et al., 2009; Zuma et al., 2009), but the effect of the hydraulic loading rate on the pollutant removal efficiency was not yet investigated.

Wastewater influent quality and hydraulic loading rate highly influence the performance of wastewater treatment filters (Healy et al., 2007). There is thus a need to optimise the pollutant removal efficiencies from the high strength grey water generated in urban slums using media based filter systems. In a recent related study, Katukiza et al. (2013b) showed that grey water from an urban slum of Bwaise III in Kampala (Uganda) has high organic, nutrient and microbial pollution loads. The specific objective of this study was therefore to optimise the filtration media (silica sand, crushed lava rock, granular activated carbon) for removal of COD, TSS, nutrients (N, P), *E. coli*, *Salmonella* spp. and total coliforms from slum grey water.

7.2   Materials and methods

7.2.1   Study area and household selection

The study area was in the Bwaise III slum, located in Kampala city (Uganda, 32° 34'E and 0° 21'N). It is not sewered and onsite sanitation in the form of elevated pit latrines is commonly used (Katukiza et al., 2010). Grey water generated at households is discharged untreated into adjacent tertiary drains or in open spaces. Ten representative households selected in a previous study (Katukiza et al., 2013b) were the source of grey water during this study. The criteria for household selection included use of both tap and spring water, equal representation of tenants and landlords using pit latrines for excreta disposal, households with and without children and the willingness to pour different grey water types in separate containers.

7.2.2   Grey water collection

Grey water was collected from households in separate containers (jerry cans) depending on the source (kitchen, washing and laundry). Collection of separate grey water types ensured that equal portions of the grey water types were supplied in the grey water influent. Grey water was collected between 8 and 10 am from the households every other day and transported to Makerere University for treatment in the filter column set-ups during the period of April 2010 to May 2012.

7.2.3   Filter column set ups

The filter columns were composed of two identical opaque uPVC columns of nominal diameter 14.6 mm, thickness 0.7 mm and length 2 m (with 60 cm depth of the media) (Figure 7.1). The filter columns were filled to the 20 cm mark with distilled water to allow uniform media packing and then manually packed to a depth of 15 cm with a gravel layer (media support with grain size 5-10 mm) followed by 10 cm intervals of media grains. Filter column 1 was packed with a 60 cm layer of crushed lava rock

(1.18-2.56 mm grain size), while filter column 2 was packed with a 30 cm layer of crushed lava rock (1.18-2.56 mm grain size), followed by a 30 cm layer of silica sand (1.18-2.56 mm grain size) at the top. The silica sand was replaced by granular activated carbon (GAC 830W) after 180 days of operating the filter columns operated in series.

7.2.4 Operating conditions

The filter columns were first operated in parallel during the ripening period and for three months after the ripening period at a constant hydraulic loading rate (HLR) of 20 cm/day. The HLR was then increased to 40 cm/day on day 135 for 60 days. From day 195 onwards, the filter columns were operated in series at a HLR of 20 cm/day and 40 cm/day. The HLR was based on previous column studies using high strength artificial wastewater (Dalahmeh et al., 2012; Logan et al., 2001), unsettled sewage (Lens et al., 1994) and olive mill wastewater (Achak et al., 2009). Liquid sampling points were located at a depth of 5 cm, 10 cm, 20 cm, 30 cm, 45 cm and 80 cm below the filter bed surface for each column (Figure 7.1).

During the ripening period, the columns were first run at a HLR of 20 cm/day until steady state conditions were attained, i.e. when the differences between successive measurements of both COD and TOC was below ±1% (Abel et al., 2012). For all operating conditions, equal volumes (5 L) of grey water types were mixed uniformly in a 20 L plastic bucket using a stirrer and then allowed to settle in the bucket for 1 hour. The settled grey water was manually poured in the influent tanks once every two days (Figure 7.1). After experimental runs in parallel mode, columns 1 and 2 were operated in series to investigate if this would increase the pollutant removal efficiency. The filter column 1 effluent was pumped into the filter column 2 influent tank using a peristaltic pump. Column 2 was then fed with filter column 1 effluent under gravity at the same HLR as filter column 1. The organic loading rate on filter column 1 was 248 g $BOD_5.m^{-2}.d^{-1}$ and 496 $BOD_5.m^{-2}.d^{-1}$, while that on column 2 was 70 g $BOD_5.m^{-2}.d^{-1}$ and 140 g $BOD_5.m^{-2}.d^{-1}$ at a HLR of 20 cm/day and 40 cm/day, respectively.

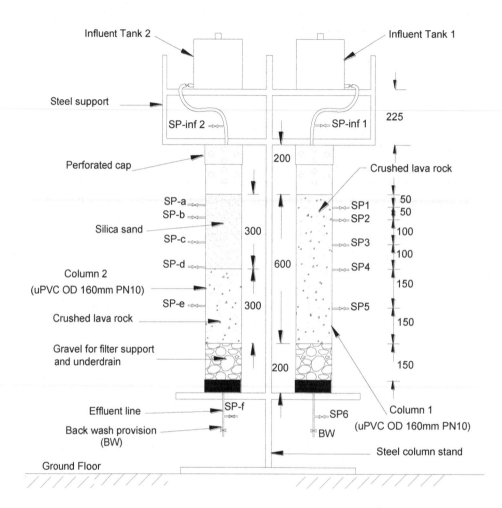

Figure 7.1: A cross-section of the column experimental set-up (SP refers to sampling point and all dimensions are in mm)

7.2.5 Characterization of the filter media

The crushed lava rock used in this study was obtained from the volcanic hills of Kisoro district (south-western Uganda) and silica sand was obtained from Gaba water treatment plant in Kampala (Uganda). The granular activated carbon (GAC) was kindly were provided by Norit Nederland BV (Amersfoort, The Netherlands). The characteristics of silica sand and crushed lava rock were determined, while those of granular activated carbon (GAC830 W) were provided by the manufacturer (Norit

Nederland BV). The properties of the crushed lava rock and silica sand that were determined included bulk density, particle density, porosity, coefficient of uniformity, coefficient of curvature and the specific surface area.

7.2.5.1 Bulk density, particle density and porosity of the crushed lava rock and silica sand

The average bulk density of the dry media was determined using a measuring cylinder and a weighing scale. It was calculated using the following expression:

$$\rho_b = \frac{m_l}{v_b}$$

.... (1)

Where ρ_b is the average bulk density in kg.m^{-3}, m_l is the mass of the dried crushed lava rock and v_b is the volume of the crushed lava rock.

The average particle density was determined using the liquid immersion method (Rühlmanna et al., 2006). 40g of crushed lave rock was added to a known volume of distilled water in a measuring cylinder and left to stand for 24 hr to allow saturation of the media. The particle density was then calculated using the following expression:

$$\rho_p = \frac{m_l}{v_p}$$

.... (2)

Where ρ_p is the average particle density in kg.m^{-3}, m_l is the mass of the crushed lava rock and v_p is the volume of the crushed lava rock particles excluding pore space.

The porosity of the crushed lava rock was then determined based on the average bulk and particle densities as follows:

$$\eta = \left(1 - \frac{\rho_b}{\rho_p}\right) x\ 100$$

.... (3)

Where η is the porosity (%).

7.2.5.2 Coefficient of uniformity and coefficient of curvature of crushed lava rock and silica sand

Sieve analysis of the crushed lava rock media was carried out according to ASTM (1998) to obtain the effective size D_{10}, D_{30} and D_{60}. The coefficient of uniformity (C_u) and coefficient of curvature (C_c) were then obtained as follows:

$$C_u = \frac{D_{60}}{D_{10}}$$

.... (4)

$$C_c = \frac{(D_{30})^2}{(D_{60} x D_{10})}$$

.... (5)

Where C_u is the coefficient of uniformity, C_c is the coefficient of curvature, and D_{10}, D_{30} and D_{60} are the sizes of the sieve through which 10%, 30% and 60% of the media would pass.

7.2.5.3 pH at the point of zero charge of the media

The pH at the point of zero charge (pH_{pzc}) was determined by the potentiometric titration method (Appel et al., 2003). 5 mg of the crushed lava rock was added to 10 mL KNO_3 solution. 0.1 M HCl or 0.1 M NaOH was then added to 0.1, 0.01 and 0.001 M KNO_3 to have the pH of each electrolyte in the pH range of 3-11. The amounts of H^+ and OH^- adsorbed by a material were determined by subtracting the amount of HCl or NaOH required to have the same pH for 10 mL of the electrolyte with no adsorbent added. The point of intersection of the titration curves (pH plotted against surface charge or amounts of acid or base added) for different concentrations of the electrolytes (varying ionic strength) was taken as pH_{pzc}.

7.2.5.4 Specific surface area of the media

The specific surface area was determined based on the Brunaeur, Emmet and Teller (BET) theory with the Micromeritics ASAP® 2020 Accelerated Surface Area and Porosimetry Analyzer that uses the gas (nitrogen) sorption technique (Kalibbala et al., 2012; Sekomo et al., 2012). The volume of nitrogen gas (at a temperature of 77K) required to form a monolayer on the sample surface enabled the calculation of the surface area of a porous solid at a range of relative pressure values (p/p_o; the ratio of the pressure of the gas and pressure at saturation) of 0.05 - 0.3. The surface area occupied by a single molecule of adsorbate on the surface was calculated from the density of the liquefied adsorbate. The specific surface area was obtained for the crushed lava rock and silica sand used in the column experiments.

7.2.6   Analytical techniques

The pH, temperature, dissolved oxygen (DO) and electrical conductivity (EC) were measured using portable WTW (Wissenschaftlich-Technische Werkstätten) meters pH 3310; Oxi 3310 and Cond 3310. The total organic carbon (TOC) and dissolved organic carbon (DOC) of non-filtered and filtered samples, respectively, were determined using a TOC-5000A (Shimadzu, Milton Keynes, UK). Samples for measurement of DOC were filtered through 0.45 μm cellulose nitrate membrane filters after leaching the filter DOC using 500 mL of distilled water in the absence of water purified with resin filters and deionization. The total suspended solids (TSS) concentration was measured following standard methods (APHA, AWWA and WEF, 2005).

Chemical parameters (COD, BOD$_5$, TP, ortho-P, TKN, NH$_4^+$-N and NO$_3^-$-N) were determined using standard methods (APHA, AWWA and WEF, 2005). COD for non-filtered samples was determined using the closed reflux colorimetric method, BOD$_5$ using the HACH BOD track (serial no. 26197-01; Loveland, Co 80539, USA). The oil and grease concentration was obtained by solid phase extraction (SPE) using USEPA Method 1664A (USEPA, 1999). Kjeldahl Nitrogen (TKN) was determined using the Kjeldahl method, whereas NH$_4^+$-N and NO$_3^-$-N were determined using, respectively, the direct nesslerization method and the calorimetric method with Nitraver 5. Total phosphorus was determined using persulfate digestion followed by the ascorbic acid spectrophotometric method, while ortho-phosphorus was measured using the ascorbic acid spectrophotometric method. The numbers of *E. coli*, total coliforms and *Salmonella* spp. were determined with Chromocult® Coliform Agar media (Merck KGaA, Germany) using the spread plate method (APHA, AWWA and WEF, 2005).

7.3   Results

7.3.1   Characteristics of silica sand, crushed lava rock and granular activated carbon

The effective size (D$_{10}$) of silica sand, crushed lava rock and GAC was 0.65 mm, 0.95 mm and 0.9 mm, while the coefficient of uniformity (C$_u$) was 4, 1.89 and 1.7, respectively (Table 7.1). GAC had a surface area (S$_{BET}$) of 1100 m^2.g^{-1}, which was higher than 0.43 m^2.g^{-1} of silica sand and 3.1 m^2.g^{-1} of crushed lava rock. In addition, the porosity of crushed lava rock was about 1.7 times that of sand (Table 7.1). The pH at zero charge (pH$_{zc}$) of silica sand and crushed lava rock was 2.5 and 7.12, respectively, while the pH of GAC was indicated as alkaline by the manufacturer (Table 7.1).

7.3.2   Characteristics of the non-settled grey water and the filter influent (settled grey water mixture)

Table 7.2 shows the physical and chemical characteristics of grey water from Bwaise III before and after settling. The dissolved oxygen (DO) of the settled grey water after aeration in the filter column influent tanks was 2.2 (±0.9) mg.L^{-1}. The COD, TSS and oil and grease concentration were reduced by 48%, 65% and 67% after settling of grey water for 1 hour in the mixing container and subsequent settling in the filter column influent tanks. The BOD_5 to COD ratio of the non-settled grey water was 0.25, while that of the settled grey water was 0.39 (Table 7.2). The settling of grey water had limited effect on the concentration of microorganisms. The concentration of total coliforms, *E. coli* and *Salmonella* spp. in settled grey water amounted to 6.9×10^7 ($\pm 2.4 \times 10^7$) cfu.(100 mL)$^{-1}$, 4.2×10^6 ($\pm 2.5 \times 10^6$) cfu.(100 mL)$^{-1}$ and 6.70×10^4 ($\pm 2.9 \times 10^4$) cfu.(100 mL)$^{-1}$, respectively (Table 7.2).

Table 7.1:   Characteristics of the filter media used in the filter column experiments treating urban slum grey water

Parameter	Unit	Parameter values of filter media used in this study			Chemical composition (%)			
		Crushed lava rock	Silica Sand	GAC	Chemical compound	Crushed lava rock	Silica sand	
Particle size range	mm	0.5 - 2.56	0.5 - 2.56	0.6 - 2.36	SiO_2	46.8	89.1	
Effective size (D_{10})	mm	0.95	0.65	0.90	Al_2O_3	15.4	3.60	
D_{30}	mm	1.60	2.00	-	Fe_2O_3	13.7	3.10	
D_{60}	mm	1.80	2.60	-	CaO	11.1	1.84	
Coeficient of uniformity C_u	-	1.89	4.00	1.70	MgO	3.7	1.20	
Cofficient of curvature C_c	-	1.42	1.54	-	TiO_2	2.9	0.09	
Ash content	% (by mass)	-	-	12	Na_2O	2.3	0.03	
Surface area, S_{BET}	$m^2.g^{-1}$	3.1	0.43	1100	K_2O	2.0	0.43	
Bulk density	$kg.m^{-3}$	1,350	1,685	470	P_2O_5	-	-	
Particle density	$kg.m^{-3}$	3650	2,640	-	BaO	0.3	-	
Porosity	%	63	36.2	-	MnO	0.3	0.05	
Ball-pan hardness	-	-	-	97				
pH_{zc}	-	7.12	2.50					
pH	-			Alkaline				

GAC refers to granular activated carbon

Table 7.2:   Characteristics of settled (column influent) and non-settled grey water

Prameter	unit	Parameter values	
		Non-settled grey water (n=30)	Settled grey water (n=30)
Temperature	$^{\circ}$C	24.3±2.5	25.0±2.5
pH	-	7.6±1.2	7.2±1.9
EC	μS.cm^{-1}	2097±135	2067±143
DO	mg.L^{-1}	1.2±0.3	2.2±0.6
COD	mg.L^{-1}	5470±1075	2861±315
BOD$_5$	mg.L^{-1}	1354±389	1125±585
TOC	mg.L^{-1}	940±161	892±124
DOC	mg.L^{-1}	568±102	559±121
TSS	mg.L^{-1}	2850±689	996±317
Oil and grease	mg.L^{-1}	21.0±6.9	5.8±2.1
NH$_4^+$-N	mg.L^{-1}	25.5±7.5	24.7±8.1
NO$_3^-$-N	mg.L^{-1}	3.1±0.6	3.8±0.5
TKN	mg.L^{-1}	64.5±15.7	58.5±9.8
Total-P	mg.L^{-1}	3.2±0.4	2.9±0.5
Ortho-P	mg.L^{-1}	2.8±0.6	2.7±1.3
Total coliforms	cfu.(100 ml)$^{-1}$	$7.5 \times 10^7 (\pm 1.3 \times 10^7)$	$6.9 \times 10^7 (\pm 2.4 \times 10^7)$
E. coli	cfu.(100 ml)$^{-1}$	$4.0 \times 10^6 (\pm 2.4 \times 10^6)$	$4.20 \times 10^6 (\pm 2.5 \times 10^6)$
Salmonella spp.	cfu.(100 ml)$^{-1}$	$8.4 \times 10^4 (\pm 4.3 \times 10^4)$	$6.70 \times 10^4 (\pm 2.9 \times 10^4)$

7.3.3 COD, TOC, DOC and TSS removal by filter columns

A stable COD and TOC removal efficiency of 70% and 73%, respectively, was achieved after about 44 days of operating the filter columns 1 and 2 in parallel at a HLR 20 cm/day during the ripening period (Figure 7.2 and Table 7.3). The DOC and TSS removal efficiencies at a HLR of 20 cm/day after the ripening period were, respectively, 69% and 85%, and 66.8% and 86% by the columns 1 and 2 (Table 7.3). When grey water was loaded on the filter columns operated in parallel without pre-treatment, the COD concentrations of the filter column 1 and 2 effluents were 1919 (±351) mg/L and 1960 (±375) mg/L, respectively (Figure 7.3A). The EC of the settled grey water (column 1 influent) was 2250 (± 125) µS/cm, while the EC of the column 2 effluent was 1144 (± 78) µS/cm (Figure 7.4).

Removal efficiencies were then determined when the columns were operated in series. The COD removal efficiency was 90% at a HLR of 20 cm/day and 84% at a 40 cm/day HLR (Figure 7.3B). The highest TSS removal efficiency of 94% was achieved at a HLR of 20 cm/day when the filter columns 1 and 2 were operated in series. TOC and DOC removal efficiencies were, respectively, 80% and 78% at a HLR of 20 cm/day and 72% and 67% at a HLR of 40 cm/day (Figure 7.5). The highest removal efficiencies of COD, TOC, DOC of 94%, 88.2% and 86.3% were achieved when the filter columns were operated in series at a HLR of 20 cm/day with GAC medium in filter column 2 (top 30 cm).

The filter columns were backwashed whenever the filtration rate reduced to less than 15% of its initial value. Filter column 1 was back-washed after every 8 weeks, while column 2 was back-washed after every 4 weeks of operation in parallel mode. In series mode, filter columns 1 and 2 were all backwashed once in two months.

Table 7.3: Removal efficiency of various parameters after ripening period during grey water treatment by filter columns in parallel

Parameter	Unit	Influent (N = 30)	Removal Efficiency (%) and log removal for microorganisms (HLR = 20 cm.d⁻¹)		Removal Efficiency (%) and log removal for microorganisms (HLR = 40 cm.d⁻¹)	
			Column 1	Column 2	Column 1	Column 2
			Crushed lava rock	Crushed lava rock and sand	Crushed lava rock	Crushed lava rock and sand
COD	mg.L⁻¹	2861±315	70±2	70±4	70±5	69±3
BOD$_5$	mg.L⁻¹	1125±585	67±7	64±6	61±2	61±5
TOC	mg.L⁻¹	892±124	71±6	70±2	66.7±2	69±8
DOC	mg.L⁻¹	559±121	69±4	66.8±5	65±4	61±5
TSS	mg.L⁻¹	996±317	85±3	86±2	80±3	79±5
NH$_4^+$-N	mg.L⁻¹	24.7±8.1	69±5	68±3	62±2	61±4
NO$_3^-$-N	mg.L⁻¹	3.8±0.5	55±2	54±3	54±2	51±2
TKN	mg.L⁻¹	58.5±9.8	51±2	43±8	42±7	39±5
Total-P	mg.L⁻¹	2.9±0.5	51±3	52±2	51±2	49±3
Ortho-P	mg.L⁻¹	2.7±1.3	51±5	48±4	48±1	48±3
Total coliforms	cfu.(100 ml)⁻¹	6.9x10⁷ (±2.4x10⁷)	1.83±0.15	1.65±0.09	1.62±0.09	1.55±0.10
E. coli	cfu.(100 ml)⁻¹	4.20x10⁶(±2.5x10⁶)	2.52±0.08	2.18±0.07	2.31±0.10	1.98±0.06
Salmonella spp.	cfu.(100 ml)⁻¹	6.70x10⁴(±2.9x10⁴)	2.25±0.04	2.37±0.08	2.10±0.03	2.07±0.1

Column 1 has a medium of crushed lava rock of depth 60 cm and size 0.5 - 2.56 mm supported by 15 cm depth of gravel.

Column 2 has media of crushed lava rock of size 1.18 - 2.36 mm (bottom depth 30 cm) and silica sand of size 1.18 - 2.36 mm (top 30 cm) supported by 15 cm depth of gravel for underdrain.

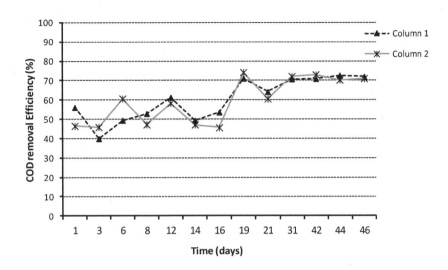

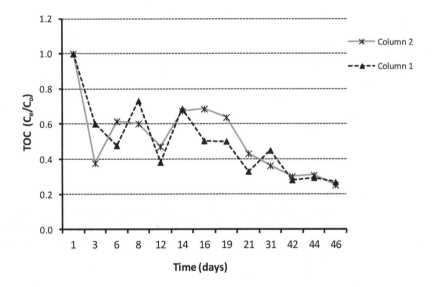

Figure 7.2: Evolution of COD and TOC removal during the filter columns' ripening period at a hydraulic loading rate of 20 cm/day. C_e and C_o refer to the TOC concentration in the column effluent and influent respectively

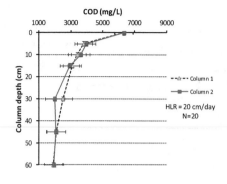

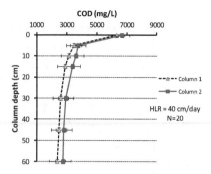

A

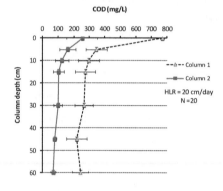

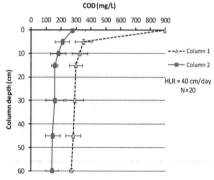

B

Figure 7.3:   COD profiles after the ripening period for column 1 (60 cm with lava rock) and column 2 (top 30 cm with sand and bottom 30 cm with lava rock) operated at a constant hydraulic loading rates of 20 cm/day and 40 cm/day A) in parallel without grey water pre-treatment and B) in series with grey water pre-treatment

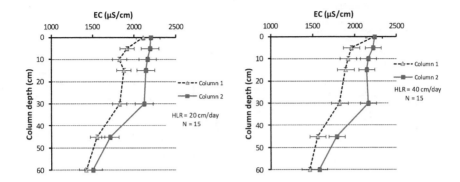

Figure 7.4: EC profiles after the ripening period for column 1 (60 cm with lava rock) and column 2 (top 30 cm with sand and bottom 30 cm with lava rock) operated in parallel at a constant hydraulic loading rates of 20 cm/day and 40 cm/day

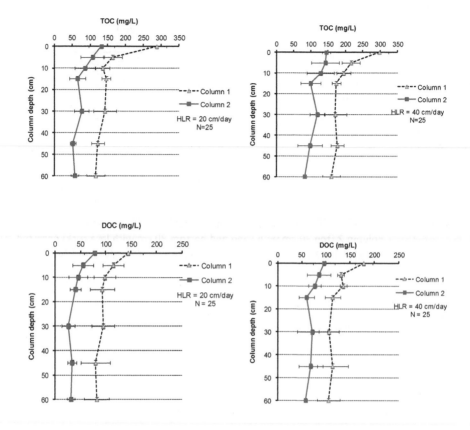

Figure 7.5: TOC and DOC profiles after ripening period for column 1 (60 cm with lava rock) and column 2 (top 30 cm with sand and bottom 30 cm with lava rock) operated in series at hydraulic loading rates of 20 cm/day and 40 cm/day

7.3.4 Nutrient removal by filter columns

The removal efficiencies of TKN, NH_4^+-N, NO_3^--N, TP and Ortho-P by the filter columns operated in series were determined at a HLR of 20 cm/day and 40 cm/day after the ripening period. The removal efficiency of TKN at a HLR of 20 cm/day and 40 cm/day was 69.1% and 65%, respectively (Figure 7.6A). The influent TKN concentration was 64.5 (± 15.7) mg.L^{-1} (Table 7.2), while the filter column 2 effluent TKN concentrations at a HLR 20 cm/day and 40 cm/day were, respectively, 19.0 mg/L and 22.0 mg/L. The highest removal efficiencies of NH_4^+-N and NO_3^--N were, respectively, 61.8% and 96% at a HLR of 20 cm/day (Figure 7.6B). The removal efficiencies of TP and Ortho-P at 20 cm/day were 86% and 84% (Figure 7.7A). When the silica sand in filter column 2 was replaced by the same depth (30 cm) of granular activated carbon, there was a complete removal of TP and Ortho-P in the grey water (Figure 7.7B). The removal efficiencies of NH_4^+-N, NO_3^--N and TKN with GAC medium were 88.3%, 97.1% and 76.2%, respectively.

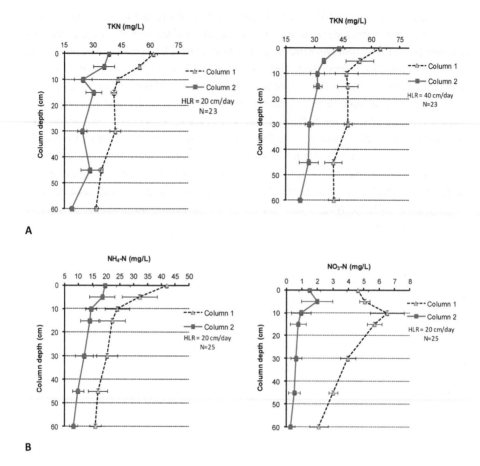

A

B

Figure 7.6: TKN, NH$_4^+$-N and NO$_3^-$-N profiles after ripening period for column 1 (60 cm with lava rock) and column 2 (top 30 cm with sand and bottom 30 cm with lava rock operated in series at constant hydraulic loading rates of 20 cm/day and 40 cm/day

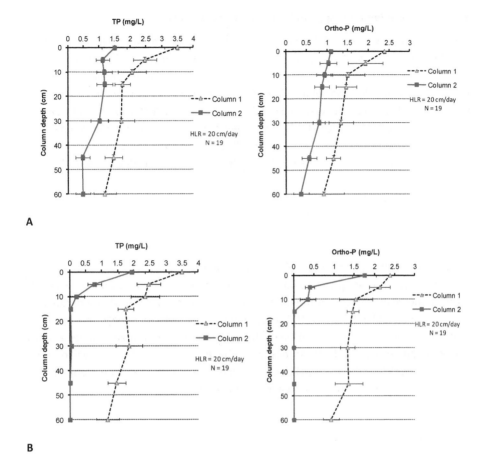

Figure 7.7: Total phosphorus (TP) and Ortho-P profiles after ripening period for A) column 1 (60 cm with lava rock) and column 2 (top 30 cm with sand and bottom 30 cm with lava rock) and B) column 1 (60 cm with lava rock) and column 2 (top 30 cm with granular activated carbon Norit GAC 830 W and bottom 30 cm with lava rock) operated at a hydraulic loading rate of 20 cm/day

7.3.5   *E. coli*, *Salmonella* spp. and total coliforms removal

The log removal of *E. coli*, *Salmonella* spp. and total coliforms at a HLR of 20 cm/day by filter column 1 was 2.52, 2.25 and 1.83, respectively, while the log removal at 40 cm/day was 2.31, 2.1 and 1.62, respectively (Table 7.3). In addition, the log removal of *E. coli*, *Salmonella* spp. and total coliforms by filter column 2 operated in parallel with filter column 1 was, respectively, 2.18, 2.37 and 1.65 at a HLR 20 cm/day and 1.98, 2.0 and 1.55 at a 40 cm/day HLR (Table 7.3). When filter columns 1 and 2 were operated in series, the log removal of *E. coli*, *Salmonella* spp. and total coliforms amounted to, respectively, 2.23, 2.13, 1.95 and 2.0, 1.27 1.98 at a HLR = 20 cm/day (Table 7.4).

The resultant log removal of *E. coli*, *Salmonella* ssp. and *total coliforms* by columns 1 and 2 in series at a HLR of 20 cm/day was 3.68, 3.50 and 3.95, respectively, while the corresponding log removal at a HLR of 40 cm/day was 3.1, 3.2 and 2.98, respectively. When silica sand in filter column 2 was replaced with GAC, the highest log removal of *E. coli*, *Salmonella* ssp. and total coliforms obtained, was 3.76, 3.63 and 3.87 at a HLR of 20 cm/day. Operating the columns in series at a HLR of 20 cm/day resulted in higher removal efficiencies of the monitored microorganisms than when operating in parallel mode (Tables 7.3 and 7.4). The log removal of *E. coli*, *Salmonella* spp. and total coliforms of more than 3 was achieved when columns were operated in series at both a HLR of 20 cm/day and 40 cm/day with the exception of the removal of total coliforms at a HLR of 40 cm/day (Table 7.4). When the filter columns were operated in parallel, the log removal of microorganisms ranged from 1.83 to 2.52.

Table 7.4: Removal of *E. coli*, *Salmonella* spp. and total coliforms from settled grey water after ripening period by filter columns in series

Column	Influent[†]			Effluent[†]			Log removal (HLR = 20 cm/day)[††]		
	E. Coli	*Salmonella* spp.	TC	*E. Coli*	*Salmonella* spp.	TC	*E. Coli*	*Salmonella* spp.	TC
1*	1.78E+07 (±1.57E+07)	3.15E+06 (±2.4E+06)	3.29E+08 (±3.51E+08)	1.5E+05 (±1.27E+05)	2.34E+04 (±1.72E+04)	3.69E+06 (±4.04E+06)	2.229	2.129	1.950
2**	3.68E+05 (±5.76E+05)	1.84E+04 (±1.43E+04)	3.45E+06 (±3.75E+06)	3.71E+03 (±3.5E+03)	9.99E+02 (±1.52E+03)	3.66E+04 (±3.26E+04)	1.996	1.265	1.975
Overall							3.681	3.498	3.954

[†]Concentrations are in cfu/100 ml and in brackets is the standard deviation. TC refers to total coliforms. The effluent of filter column 1 was pumped to the influent tank of filter column 2

[††]The log removal of *E. coli*, *Salmonella* spp. and total coliforms at a hydraulic loading rate (HLR) of 40 cm/day by columns 1 and 2 in series was 3.1, 3.2 and 2.98, respectively.

*Filter media of crushed lava rock of depth 60 cm.

**Filter media consists of top 30 cm depth silica sand and bottom 30 cm depth of crushed lava rock.

7.4 Discussion

7.4.2 Grey water characteristics

The COD and BOD_5 concentrations of, respectively, 5470 (±1075) mg/ and 1354 (±389) mg/L (Table 7.2) show that grey water from Bwaise III is a high strength grey water. The BOD_5 to COD ratio of 0.25 also indicates that it is not easily biodegradable (Li et al., 2009; Metcalf and Eddy, 2003) and thus needs to be treated by a combination of both straining and biological treatment processes. This makes media (sand, gravel, lava) filter based systems appropriate for COD removal from grey water.

The high concentration of TSS (> 2500 mg/L) of grey water from Bwaise III as a result of multiple uses of water contributes to the low biodegradability of grey water. The low BOD_5 to COD ratio of grey water from peri-urban areas ranging from 0.21 to 0.4 has also been reported in previous studies (Kulabako et al., 2011; Sall and Takahashi, 2006). Grey water from Bwaise III also contains oil, grease and TSS (Table 7.2), which can potentially cause filter clogging and reduce filter performance. The raw grey water was therefore settled to reduce the TSS concentration by 65%, while the top layer was also skimmed off, resulting in a reduction of 72% of the oil and grease concentration (Table 7.2). Oil and grease also interferes with the biological, physical and chemical processes during wastewater treatment (Cammarota and Freire, 2006; Travis et al., 2008). Therefore, pre-treatment of grey water to remove TSS and oil and grease is important to prevent filter clogging and reduced performance that would eventually result in bad odours. In practice, the settled solids may be dewatered on drying beds with an under-drain prior to use as a soil conditioner in the back yard garden, while the leachate may be mixed with the grey water for treatment.

Grey water from Bwaise III has a high TKN (> 50 mg/L) and TP (3.2 ± 0.4 mg/L) concentration (Table 7.2). The phosphorous concentration may be higher in grey water where phosphorus is used as an ingredient in production of detergents (Li et al., 2009; Metcalf and Eddy, 2003). The concentration of NH_4^+-N of 25.5 ± 7.5 mg/L in the feed grey water indicates contamination from urine. The bacteriological quality of the grey water from Bwaise III slum was comparable to that of raw sewage. The 10^4 to 10^7 cfu/100 mL concentration of E. coli, Salmonella spp. and total coliforms (Table 7.2) in the grey water from Bwaise III was comparable to typical concentrations of these microorganisms in sewage (Metcalf and Eddy, 2003). The concentration of E. coli in grey water from peri-urban areas ranging from 5.8 x 10^3 cfu/100 mL to 1.5 x 10^7 cfu/100 mL and even higher concentrations of total coliforms have been reported (Carden et al., 2007; Kulabako et al., 2011; Sall and Takahashi, 2006). Grey water from urban slums thus poses a risk to the health of the slum dwellers and requires treatment before disposal or reuse.

7.4.3 The role of pre-treatment

This study showed that pre-treatment of grey water is essential to prevent clogging of media based filters and sustain the performance. Clogging of media (sand, gravel) based filters is caused by accumulation of the suspended solids and oil and grease present in the grey water and biofilm formation on the media surface (Cammarota and Freire, 2006; Leverenz et al., 2009. In addition, organic loading rates exceeding than 22 g $BOD_5.m^{-2}.d^{-1}$ at a HLR ranging from 12 to 20 $cm.d^{-1}$ and an increase in TSS loading rate accelerated the clogging and reduced the performance of intermittent sand filters (Healy et al., 2007). In this study, clogging of filter columns was minimised by the sedimentation of raw grey water as indicated by the reduction of TSS, and oil and grease concentration (Table 7.2). A longer hydraulic retention to reduce influent TSS and COD concentration after grey water pre-treatment was also provided by further grey water settling in the influent tanks. The rate of clogging of the sand was two times that of crushed lava rock. This can be attributed to the differences in porosity and surface area (Table 7.1). The rate of clogging may be reduced by scrapping of the top layer to increase the filtration rate (Healy et al., 2007; Rodgers et al., 2004). Moreover, in practice, the wetting and drying cycles as a result of intermittent flow of grey water through the filter system under conditions in the slums reduces clogging (Essandoh et al., 2013; Kadam et al., 2009).

7.4.4 Performance of the filter columns in parallel and in series

The pollutant removal efficiencies during the filter column experiments show that media (silica sand, crushed lava rock) filters have the potential to reduce the grey water pollutant loads originating from households in typical urban slums. The highest removal efficiency of COD and TSS exceeding 90% was achieved with filter columns 1 and 2 in series at a HLR of 20 cm/day. At a HLR of 40 cm/day, the COD removal fell back by 6% to 10%. In addition, the COD removal efficiency by filter columns 1 and 2 in parallel at HLR of 20 cm/day and 40 cm/day was also 10 - 20% lower than the when filter columns were operated in series. Physical straining had a key role in the COD removal because of the low COD to BOD_5 ratio (Table 7.2). More than 60% of the COD removal occurred in the top 15 cm of the infiltration depth (Figure 7.3). Indeed, the highest pollutant removal efficiency in the media (gravel and sand) filters occurs in the top 10 cm of the infiltration depth (Campos et al., 2002; Rodgers et al., 2005). This can be partly attributed to the retention of the particulate matter in the filter. In addition, an increase in the COD removal efficiency during the ripening period of about 20% (Figure 7.2) shows that biological activity in the top layer also contributed to the removal of the biodegradable organic matter.

The dissolved Kjeldahl nitrogen and phosphorus removal was mainly attributed to precipitation, while their removal by adsorption may have been limited as a result of the high pH (> 7) of grey water. The replacement of silica sand by granular activated carbon (GAC) resulted in complete removal of phosphorus, which was attributed to adsorption and precipitation (Vohla et al., 2007). Moreover, the GAC had a higher specific surface area than crushed lava rock and silica sand (Table 7.1). The application of GAC as a filter medium in slums is unlikely because of its high cost and not being locally available, despite its higher efficiency in removal of phosphorus compared to silica sand and crushed lava rock. Particulate Kjeldahl nitrogen removal may be linked to the high removal of COD and TSS in the top 15 cm of the infiltration depth. About 20 to 35% of TKN concentration in the influent occurred after 15 cm of grey water infiltration, which is high when compared to TKN removal efficiency of 65% and 69% by filter columns 1 and 2 in series at 20 cm/day and 40 cm/day.

The removal of NH_4^+-N was probably due to adsorption by the media, especially by crushed lava rock with a large surface area and adsorptive properties compared to silica sand. There was a reduction of 61.8% and 57% at 20 cm/day, while there was no major increase in the concentration of NO_3^--N (Figure 7.6B) in the series of filter columns 1 and 2. Nitrogen removal by nitrification and denitrification was thus limited because of the low oxygen concentration in the filter column influent (Table 7.2). Nitrification could be enhanced by artificial aeration with compressed air, rather than natural aeration through openings as was the case during this study (Dalahmeh et al., 2012). Loss of NH_4^+ by volatilisation was unlikely because the pH of the filter columns 1 and 2 effluents ranged from 6 to 6.9, which is below the pH 8 required for conversion of NH_4^+ to NH_3 (Hammer and Knight, 1994). The removal efficiencies of TKN, NH_4^+-N and NO_3^--N by filter columns 1 and 2 in series were reduced by 4% to 6% when the HLR was increased from 20 to 40 cm/day. In addition, operating the filter columns in series resulted in increased nutrient removal efficiencies of TKN, NH_4^+-N and NO_3^--N (10-18%, 12-20% and 30-40%, respectively).

The removal of *E. coli*, *Salmonella* spp. and total coliforms during filtration was attributed to different mechanisms, including biological activity (predation, die-off), exposure to sunlight as well as retention on the filter media by straining and adsorption (Auset et al., 2005; Bomo et al., 2004; Campos et al., 2002) in the extended infiltration depth of columns 1 and 2. In addition, removal efficiencies of microorganisms show different dependencies on the hydraulic loading rate according to their size and sticking efficiencies (Auset et al., 2005; Campos et al., 2002; Lutterodt et al., 2009). Therefore, the removal of smaller size pathogens such as viruses under similar conditions requires an investigation in case of reuse of the final effluent subject to guidelines such as WHO guideline for irrigation. Protozoa and helminths are effectively removed by media (sand, soil, lava) filters (Bomo et al., 2004; Hijnen et al., 2004), although they were not investigated in this study.

Operating filter columns 1 and 2 in series at a HLR of 20 cm/day resulted in higher pollutant removal efficiencies as a result of the extended infiltration depth and two unsaturated infiltration zones. In practice, the first filter may be used for coarse filtration of grey water, while the second filter would be operated in series using smaller medium size grains for secondary filtration. Such a system would be suitable for decentralised treatment of high strength grey water to reduce the organic load (Figures 7.3 and 7.5), nutrient load (Figures 7.6 and 7.7) and pathogen load (Tables 7.3 and 7.4) pollution load from urban slums such as Bwaise III. However, the support and incentives of the local authorities to prioritise grey water interventions in urban slums in order is essential to have a wider application of such filter systems in slums. In addition, acceptability by the users and making the filters affordable may pose a challenge to their sustainability. Generally, decentralised treatment of grey water in urban slums using such low-technology solutions that occupy less space is preferred to other technologies such as wetlands and oxidation ponds, which require availability of a large land area.

7.5 Conclusions

- Sand and crushed lava rock based filters can achieve above 85% removal of organic matter from high strength grey water and 50% to 70% removal of the nutrients nitrogen and phosphorous.
- The highest pollutant removal efficiencies were achieved in the top 15 cm of infiltration depth under all filter columns operating conditions.
- Grey water treatment with filter columns 1 and 2 operated in series resulted in higher pollutant removal efficiencies than using either filter column 1 or filter column 2 only. A roughing filtration step is, therefore, important to achieve a better grey water effluent quality.
- The log removal of *E. coli*, *Salmonella* spp. and total coliforms increased with the infiltration depth during grey water treatment with media filtration. In addition, the log removal of these microorganisms was higher when using crushed lava rock than silica sand as a filtration medium.

References

Abel, C.D.T, Sharma, S.K., Malolo, Y.N., Maeng, S.K., Kennedy, M.D., Amy, G.L., 2012. Attenuation of Bulk Organic Matter, Nutrients (N and P), and Pathogen Indicators During Soil Passage: Effect of Temperature and Redox Conditions in Simulated Soil Aquifer Treatment (SAT). Water, Air and Soil Pollution 223(8), 5205-5220.

Abu Ghunmi, L., Zeeman, G., van Lier, J., Fayyed, M., 2008. Quantitative and qualitative characteristics of grey water for reuse requirements and treatment alternatives: the case of Jordan. Water Science and Technology 58(7), 1385-1396.

Achak, M., Mandi, L., Ouazzani, N., 2009. Removal of organic pollutants and nutrients from olive mill wastewater by a sand filter. Journal of Environmental Management 90(8), 2771–2779.

Alder, A., Bruchet, A., Carballa, M., Clara, M., Joss, A., Loe_er, D., McArdell, C., Miksch, K., Omil, F., Tuhkanen, T., Ternes, T., 2006. Consumption and occurrence. In: Ternes, T., Joss, A. (Eds.), Human Pharmaceuticals, Hormones and Fragrances: The challenge of micropollutants in urban water management. IWA publishing, London.

Al-Hamaiedeh, H., Bino, M., 2010. Effect of treated grey water reuse in irrigation on soil and plants. Desalination 256(1-3), 115–119.

APHA, AWWA and WEF., 2005. Standard Methods for the Examination of Water and Wastewater. *American Public Health Association, American Water Works Association and Water Environment Federation publication,* 21st edition. Washington DC, USA.

ASTM., 1998. Manual on test sieving methods: guidelines for establishing sieve analysis procedures. Lawrence R. P., Charles, W. W. (Ed.), West Conshohocken: American Society for Testing and Materials.

Auset, M., Keller, A.A., Brissaud, F., Lazarova, V., 2005. Intermittent filtration of bacteria and colloids in porous media. Water Resources Research 41(9), W09408:1-13.

Bomo, A.M., Stevik, T.K., Hovi, I., Hanssen, J.F., 2004. Bacterial removal and protozoan grazing in biological sand filters. Journal of Environmental Quality 33(3), 1041-7.

Cammarota, M.C., Freire, D.M.G., 2006. A review on hydrolytic enzymes in the treatment of wastewater with high oil and grease content. Bioresource Technology 97, 2195-2210.

Campos, L.C., Su, M.F.J., Graham, N.J.D., Smith, S.R., 2002. Biomass development in slow sand filters. Water Research 36(18), 4543-4551.

Carden, K., Armitage, N., Winter, K., Sichone, O., Rivett, U., Kahonde, J., 2007. The use and disposal of greywater in the non-sewered areas of South Africa: Part 1- Quantifying the greywater generated and assessing its quality. Water SA 33, 425-432. ISSN 0378-4738.

Dalahmeh, S.S., Pell, M., Vinnerås, B., Hylander, L.D., Öborn, I., Håkan Jönsson, H, 2012. Efficiency of bark, activated charcoal, foam and sand filters in reducing pollutants from greywater. Water Air and Soil Pollution 223(7), 3657–3671.

Eriksson, E., Aufarth, K., Henze, M., Ledin, A., 2002. Characteristics of grey wastewater. Urban Water 4, 85-104.

Essandoh, H.M.K., Tizaoui, C., Mohamed, M.H.A., 2013. Removal of dissolved organic carbon and nitrogen during simulated soil aquifer treatment. Water Research 47, 3559-3572.

Garland, J.L., Levine, L.H., Yorio, N.C., Adams, J.L. and Cook, K.L., 2000. Graywater processing in recirculating hydroponic systems: Phytotoxicity, surfactant degradation, and bacterial dynamics.Water Research 34, 3075–3086.

Gross, A., Azulai, N., Oron, G., Ronen, Z., Arnold, M., Nejidat, A., 2005. Environmental impact and health risks associated with greywater irrigation: a case study. Water Science & Technology 52(8), 161–169.

Healy, M.G., Rodgers, M., Mulqueen, J., 2007. Treatment of dairy wastewater using constructed wetlands and intermittent sand filters. Bioresource Technology 98(12), 2268-2281.

Hernández-Leal, L., Temmink, H., Zeeman, G., Buisman, C.J.N., 2011. Removal of micropollutants from aerobically treated grey water via ozone and activated carbon. Water Research 45(9), 2887-2896.

Hijnen, W.A., Schijven, J.F., Bonné, P., Visser, A., Medema, G.J., 2004. Elimination of viruses, bacteria and protozoan oocysts by slow sand filtration. Water Science and Technology 50(1), 147-54.

Jefferson, B., Burgess, J.E., Pichon, A., Harkness, J., Judd, S.J., 2001. Nutrient addition to enhance biological treatment of greywater. Water Research 35 (11), 2702–2710.

Kadam, A.M., Nemade, P.D., Oza., G.H., Shankar, H.S., 2009. Treatment of municipal wastewater using laterite-based constructed soil filter. Ecological Engineering 35(7), 1051-1061.

Kalibbala, H.M., Wahlberg, O., Plaza, E., 2012. Horizontal flow filtration: impact on removal of natural organic matter and iron co-existing in water source. Separation Science and Technology 47, 1628-1637.

Katukiza, A.Y., Ronteltap, M., Niwagaba, C., Kansiime, F., Lens, P.N.L., 2013b. Grey water characterisation and pollution load in an urban slum. Internation Journal of Environmental Science and Technology. "Accepted article".

Katukiza, A.Y., Temanu H, Chung, J.W., Foppen, J.W.A., Lens, P.N.L., 2013a. Genomic copy concentrations of selected waterborne viruses in a slum environment in Kampala, Uganda. Journal of Water and Health 11(2), 358-369.

Kulabako, N. R., Ssonko, N.K.M, Kinobe, J., 2011. Greywater Characteristics and Reuse in Tower Gardens in Peri-Urban Areas – Experiences of Kawaala, Kampala, Uganda. The Open Environmental Engineering Journal 4, 147-154.

Labite, H., Lunani, I., van der Steen, P., Vairavamoorthy, K., Drechsel, P., Lens, P., 2010. Quantitative Microbial Risk Analysis to evaluate health effects of interventions in the urban water system of Accra, Ghana. Journal of Water and Health 8(3), 417-430.

Lens, P.N., Vochten, P.N., Speleers, L., Verstraete, W.H., 1994. Direct treatment of domestic wastewater by percolation over peat, bark and wood chips. Water Research 28(1), 17-26.

Leverenz, H.L., Tchobanoglous, G., Darby, J.L., 2009. Clogging in intermittently dosed sand filters used for wastewater treatment. Water Research 43(3), 695-705

Li, F., Wichmann, K, Otterpohl, R., 2009. Review of technological approaches for grey water treatment and reuses. Science of the Total Environment 407(11), 3439-49.

Logan, A.J., Stevik, T.K., Siegrist, R.L., Rønn, R.M., 2001. Transport and fate of *Cryptosporidium parvum* oocysts in intermittent sand filters. Water Research 35(18), 4359-4369

López-Zavala, M.A., 2007. Treatment of lower load graywater by using a controlled soil natural treatment system. Water Science and Technology 55(7), 39-45.

Lutterodt, G., Basnet, M., Foppen, J.W.A., Uhlenbrook, S., 2009. Effects of surface characteristics on the transport of multiple Escherichia coli isolates in large scale column of quartz sand. Water Research Vol. 43 p. 595-604.

Mandal. D., Labhasetwarb, P., Dhonea, S., Ajay Shankar Dubeya, A.S., Shindec, G., Watea, S., 2011. Water conservation due to greywater treatment and reuse in urban setting with specific context to developing countries. Resources, Conservation and Recycling 55(3), 356–361.

Metcalf, Eddy. Wastewater engineering: treatment and reuse. 4th ed. 1221 Avenue of Americas, New York, NY 10020, USA: McGraw-Hill Companies; 2003.

Morel, A., Diener, S., 2006. Greywater Management in Low and Middle-Income Countries. Review of different treatment systems for households or neighbourhoods. [Online] http://www.eawag.ch/organisation/abteilungen/sandec/publikationen/publicatio ns_ewm/downloads_ewm/Morel_Diener_Greywater_2006.pdf [2010, Sept.11]

Rodda, N., Carden, K., Armitage, N., du Plessis, H.M., 2011. Development of guidance for sustainable irrigation use of greywater in gardens and small-scale agriculture in South Africa. Water SA 37(5), 727-738.

Rodgers, M., Mulqueen, J., Healy, M.G., 2004. Surface clogging in an intermittent stratified sand filter. Soil Science Society of America Journal 68, 1827-1832.

Sall, O., Takahashi, Y., 2006. Physical, chemical and biological characteristics ofstored greywater from unsewered suburban Dakar in Senegal. Urban Water Journal 3(3), 153-164.

Tandlicha, R., Zuma, B.M., Whittington-Jonesc, K.J., Burgess, J.E., 2009. Mulch tower treatment system for greywater reuse Part II: destructive testing and effluent treatment. Desalination 242, 57-69.

Temmink, H., Klapwijk, B., 2004. Fate of linear alkylbenzene sulfonate (LAS) in activated sludge plants. Water Research 38 (4), 903-912.

Travis, M.J., Weisbrod, N., Gross, A., 2008. Accumulation of oil and grease in soils irrigated with greywater and their potential role in soil water repellency. Science of the Total Environment 394, 68-74.

USEPA., 2009. Method 1664 revision A: N-hexane extractable material (HEM; oil and grease) and silica gel treated N-hexane extractable material (SGT-HEM; non-polar material) by extraction and gravimetry. Washington DC: United States Environmental Protection Agency.

Vohla, C., Alas, R., Nurk, K., Baatz, S., Mander, Ü., 2007. Dynamics of phosphorus, nitrogen and carbon removal in a horizontal subsurface flow constructed wetland. Science of the Total Environment 380(1-3), 66–74.

Ying, G.G., 2006. Fate, behaviour and effects of surfactants and their degradation products in the environment. Environment International 32 (3), 417-431.

Zuma, B.M., Tandlich, R., Whittington-Jones, K.J., Burgess, J.E., 2009. Mulch tower treatment system; Part I: overall performance in grey water treatment. Desalination 242, 38-56.

Chapter 8: **A two-step crushed lava rock filter unit for grey water treatment at household level in an urban slum**

This Chapter is based on:

Katukiza, A.Y., Ronteltap, M., Niwagaba, C., Kansiime, F., Lens, P.N.L., 2013. A two-step crushed lava rock filter unit for grey water treatment at household level in an urban slum. *Journal of Environmental Management*. "Accepted".

Abstract

Decentralised grey water treatment in urban slums using low-cost and robust technologies offers opportunities to minimise public health risks and to reduce environmental pollution caused by the highly polluted grey water i.e. with a COD and N concentration of 3000-6000 mg.L^{-1} and 30-40 mg.L^{-1}, respectively. However, there has been very limited action research to reduce the pollution load from uncontrolled grey water discharge by households in urban slums. This study was therefore carried out to investigate the potential of a two-step filtration process to reduce the grey water pollution load in an urban slum using a crushed lava rock filter, to determine the main filter design and operation parameters and the effect of intermittent flow on the grey water effluent quality. A two-step crushed lava rock filter unit was designed and implemented for use by a household in the Bwaise III slum in Kampala city (Uganda). It was monitored at a varying hydraulic loading rate (HLR) of 0.5 m.d^{-1} to 1.1 m.d^{-1} as well as at a constant HLR of 0.39 m.d^{-1}. The removal efficiencies of COD, TP and TKN were, respectively, 85.9%, 58% and 65.5% under a varying HLR and 90.5%, 59.5% and 69%, when operating at a constant HLR regime. In addition, the log removal of *E. coli*, *Salmonella* spp. and total coliforms were, respectively, 3.8, 3.2 and 3.9 under the varying HLR and 3.9, 3.5 and 3.9 at a constant HLR. The results show that the use of a two-step filtration process as well as a lower constant HLR increased the pollutant removal efficiencies. Further research is needed to investigate the feasibility of adding a tertiary treatment step to increase the nutrients and microorganisms removal from grey water.

8.1   Introduction

Grey water is one of the domestic waste streams. In urban slums, it accounts for 65 % to 90% of the domestic wastewater production (Kulabako et al., 2011; Abu Ghunmi et al., 2008; Jamrah et al., 2008; Carden et al., 2007a). Grey water is mainly disposed of in the existing storm water drains and open spaces without treatment because there is no provision for its treatment. Yet, grey water from slums has a high concentration of COD (>2000 mgL^{-1}), *E. coli* (10^4-10^7 cfu.(100 mL)$^{-1}$), total phosphorus (5 - 240 mgL^{-1}), total nitrogen (5 - 200 mgL^{-1}) and heavy metals (Carden et al., 2007b; Kariuki et al., 2012; Katukiza et al., 2013a; Kulabako et al., 2010;). Grey water treatment in urban slums is therefore driven by the need to reduce environmental pollution and to minimise risks to human health from the grey water pollution load.

Discharge of untreated grey water may lead to pollution of ground water sources by nutrients and micro-pollutants, eutrophication of surface water bodies and soil salinity (Gross et al., 2005; Morel and Diener, 2006; Nyenje et al., 2010). Total suspended solids (TSS) in grey water contribute to the sedimentation and reduction of the hydraulic capacity of drainage channels and surface water reservoirs, and to clogging of media based filter systems (Leverenz et al., 2009). In addition, oil and grease in grey water reduce the soil's ability to transmit water and the grey water treatment efficiency by interfering with the biological, physical and chemical processes (Cammarota and Freire, 2006; Travis et al., 2008). Lastly, grey water contains waterborne viruses, bacteria, parasitic protozoa and helminths (Birks and Hills, 2007; Katukiza et al., 2013b; Ottoson and Stenström, 2003), which potentially cause diseases. Hence, for public and environmental health, improved grey water collection and treatment is vital

Primary, secondary or tertiary treatment of grey water may be achieved by a combination of technologies depending on the discharge and reuse requirements. They include settling tanks, septic tanks, media based (soil, sand, volcanic lava, mulch) filters, constructed wetlands, and oxidation ponds (Dalahmeh et al., 2011; Li et al., 2009; Morel and Diener, 2006; von Sperling and Chernicharo, 2005). Primary treatment of grey water removes oil and grease and reduces the TSS concentration, while tertiary treatment of grey water removes remaining biodegradable and non-biodegradable organics, pathogens, nutrients, and micro-pollutants after secondary treatment (Campos et al., 2002; Metcalf and Eddy, 2003). Therefore, the quality of grey water influences the choice for the grey water treatment technology.

A number of low-cost grey water technologies that include horizontal and vertical flow constructed wetlands, grey water tower gardens, and infiltration trenches have been applied in many parts of the world (Li et al., 2009; Masi et al., 2010; Morel and Diener, 2006). However, the existing grey water treatment technologies have neither been adapted to slum conditions nor implemented at the pollution sources (households) for effective control of the pollution load. Moreover, active participation of the slum inhabitants is important for the sustainability of a low-cost household grey water treatment system. There is thus a need for action research to demonstrate the practical application of a media filter unit for grey water treatment at household level in an urban slum. The objective of this study was, therefore, to investigate the potential of a low-cost and robust lava rock filter based on a two-step filtration process to reduce the grey water pollution load in an urban slum. The specific objectives were to implement a crushed lava rock grey water treatment filter unit operated at household level in Bwaise III in Kampala city (Uganda); to determine the removal efficiency of COD, TSS, nutrients (N, P), *E. coli*, *Salmonella* spp. and total coliforms; and to determine the main filter design and operation parameters as well as the effect of intermittent flow on the grey water effluent quality.

8.2   Materials and methods

8.2.1   Study area

The Bwaise III slum in Kampala city (Uganda) was selected as the study area for the implementation and monitoring of the performance of the grey water filter unit. It is a typical slum area located in a reclaimed wetland (32° 34′E and 0° 21′N) at an altitude of 1,170 m above sea level. There are two dry season periods in the area from January to March and June to August, while during the rainy seasons flooding is usually a problem. Bwaise III is drained by two major open storm water channels into which smaller drains that convey storm water and grey water discharge. Bwaise III is not sewered and the majority of residents use onsite sanitation in the form of elevated pit latrines. There is no grey water management system in place and residents discharge the grey water in nearby open spaces and storm water drains. The water supply sources in the area include contaminated springs and a piped water system that serves the residents who can afford to pay for tap water.

8.2.2   Household selection

The selection of the household where the grey water treatment filter was implemented in Bwaise III was critical in this study. The majority of the residents (>60%) in the study area are tenants (Isunju et al., 2013). A household was chosen among the 10 households who participated in an earlier study (Katukiza et al., 2013a). The criteria included: availability of space (about 1 m^2 of land), a household with a resident landlord or landlady to ensure the safety of the filter and the presence of the same people throughout the entire study period, a household size of at least 7 (average size in Bwaise III) with children (< 3 years) and adults, a per capita water consumption of at least 18 L.c^{-1}.d^{-1} (average value for Bwaise; Katukiza et al., 2010), the presence of a tertiary drain nearby to which the effluent would be discharged and the willingness of the household members to use the grey water treatment filter.

8.2.3 Design and implementation of the crushed lava rock filter in Bwaise III

8.2.3.1 Design

A pilot onsite grey water treatment unit using crushed lava rock was designed and implemented at a household in Bwaise III parish. The filter unit was composed of two identical filters (R1 and R2 in series) made of plastic material to avoid rusting and a filter support from hollow steel sections of 1.5 mm thickness with a concrete foundation (Figure 8.1). The filters were composed of 10 cm of the under-drain of crushed gravel (media size: 5-10 mm), 30 cm of graded crushed lava rock (media size: 2.56 - 5 mm for the first filter and 1.18-2.56 mm for the second filter; Figure 8.1), 170 cm clear space above the media and a perforated plastic diffuser. Sampling points SP1, SP2 and SP3 were provided for the influent, R1 effluent and R2 effluent, respectively.

Crushed lava rock medium was chosen as a filtration medium because it is widely available in south western Uganda in large quantities, has a higher specific surface area and porosity compared to sand and gravel, and its chemical characteristics enhance its suitability as a medium in filter systems for waste water treatment. Crushed lava rock has a specific surface area of about 6-8 times that of quartz and gravel (Kalibbala et al., 2012).

The filter was designed in such a way that at least three 200 mL samples can be obtained at anytime regardless of the usage by the household. This was made possible by raising the outlet pipes of both reactors by 10 cm and inserting sampling points about 2 to 3 cm from the reactor bottom. The elevated outlet allowed the gravel (filter media support) to remain saturated during the filter operation under intermittent flow. A perforated plastic diffuser (with holes about 2 mm diameter) was introduced on top of the reactors to distribute the filter influent uniformly on the media and to prevent scouring of the top biologically active layer. In addition, a perforated Tee- junction with end caps was provided at the end of the outlet pipe from reactor 1 to distribute the effluent of R1 on the diffuser of R 2 (Figure 8.1).

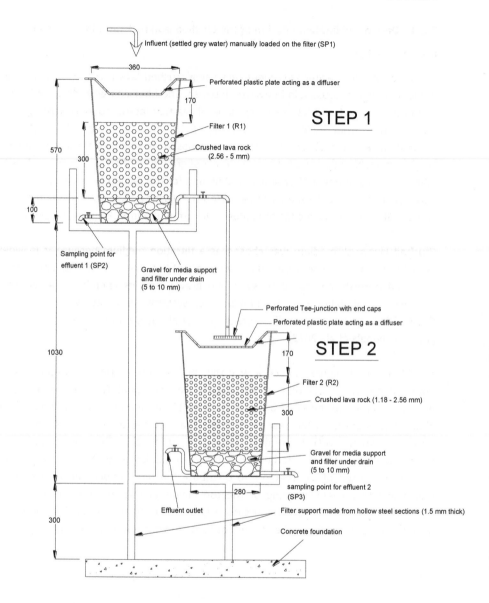

Figure 8.1: A cross-section of a crushed lava rock filter implemented at household level in Bwaise III slum in Kampala city (Uganda) (all dimensions are in mm)

A two-step design was chosen so that R1 can provide the primary treatment of settled grey water and R2 filled with a smaller grain size of crushed lava can provide secondary treatment. In addition, the highest pollutant removal in the top 10 cm of the infiltration depth for intermittent filters (Campos et al., 2002; Jellison et al., 2000; Rodgers et al., 2005) and thus two separate filters provide a longer aerobic zone than one filter with an equivalent infiltration depth. The filter was designed to serve an average household size of 7 in Bwaise III. The filter design hydraulic loading rate was 1.1 m.d^{-1} based on the grey water production in Bwaise III of about 16 L.c^{-1}.d^{-1} and the filter surface area of 0.102 m^2.

8.2.3.2 Operational conditions

The crushed lava rock filter was implemented at a household and operated under uncontrolled and then controlled intermittent flow conditions. Under uncontrolled conditions, the hydraulic loading rate varied from 0.5 to 1.1 m.d^{-1} dependent on the household grey water production, while the filtration rate after feeding R1 with settled grey water varied from 0.5 to 1.1 L.min^{-1}. For controlled conditions, 40 L.d^{-1} was loaded on the filter unit once a day in the morning (before 10 a.m.), which is equivalent to a hydraulic loading rate of 0.39 m.d^{-1}. Controlled conditions were to investigate the pollutant removal efficiencies at a constant HLR lower than the varying HLR and independent of the grey water production rate at the household.

The grey water treatment steps include sedimentation (pre-treatment) and a two-step filtration (main treatment using filter R1 and R2 in series) for all operational conditions. Sedimentation of grey water from households was done by pouring the grey water in 20 litre buckets and leaving it for 1 hour to allow for settling of the total suspended solids. The top layer of oil and grease layer and scum (usually less than 1 litre) was skimmed off by the filter users and discharge it with the settled solids. The supernatant (settled grey water) was then poured in a plastic bucket and applied on R1 (Figure 8.2). The solids were poured on a perforated plastic support on top of a separate plastic bucket for dewatering prior to use as a soil conditioner in the back yard garden, while the leachate was mixed with the supernatant.

A   **B**

Figure 8.2:   A crushed lava rock filter unit (A) implemented at a household near a tertiary drain to which the effluent is discharged (B) Grey water being poured into the filter unit, while samples are being collected

8.2.4   Characterization of the crushed lava rock

8.2.4.1   Bulk density, particle density and porosity of the media

The grey coloured medium (crushed lava rock) used in this study was obtained from the volcanic hills of Kisoro district (south-western Uganda). The average bulk density of the dry medium was determined using a measuring cylinder and a weighing scale. It was calculated using the following expression:

$$\rho_b = \frac{m_l}{v_b}$$

.... (1)

Where ρ_b is the average bulk density in kg.m^{-3}, m_l is the mass of the dried crushed lava rock and v_b is the volume of the crushed lava rock.

The average particle density was determined using the liquid immersion method (Rühlmanna et al., 2006): 40g of crushed lava rock was added to a known volume of distilled water in a measuring cylinder and left to stand for 24 h to allow saturation of the medium. The particle density was then calculated using the following expression:

$$\rho_p = \frac{m_l}{v_p}$$

.... (2)

Where ρ_p is the average particle density in kg.m^{-3}, m_l is the mass of the crushed lava rock and v_p is the volume of the crushed lava rock particles excluding pore space.

The porosity of the crushed lava rock was then determined based on the average bulk and particle densities as follows:

$$\eta = \left(1 - \frac{\rho_b}{\rho_p}\right) x\ 100$$

.... (3)

Where η is the porosity (%).

8.2.4.2 Coefficient of uniformity and coefficient of curvature of the medium

Sieve analysis of the crushed lava rock media was carried out according to ASTM (1998) to obtain the effective size D_{10}, D_{30} and D_{60}. The coefficient of uniformity (C_u) and coefficient of curvature (C_c) were then obtained as follows:

$$C_u = \frac{D_{60}}{D_{10}}$$

.... (4)

$$C_c = \frac{(D_{30})^2}{(D_{60} x D_{10})}$$

.... (5)

Where C_u is the coefficient of uniformity, C_c is the coefficient of curvature, and D_{10}, D_{30} and D_{60} are the sizes of the sieve through which 10%, 30% and 60% of the medium passed.

8.2.4.3 Chemical composition of the medium

X–ray fluorescence (XRF) spectrometry was used to determine the chemical composition of the lava rock (Kalibbala et al., 2012; Sekomo et al., 2012). The samples were analysed using a Philips PW2400 WD-XRF spectrometer and quantitative data were obtained with the uniquant program. The elemental analysis was carried out using ICP-Mass Spectrometry. 0.5 g of crushed lava rock with a particle size less than 300 μm was mixed with 10 mL of concentrated nitric acid in a digestion tube. The digestion was carried out in a Microwave Accelerated Reaction System (CEM Mars 5) at a pressure of 10 bars. Appropriate dilutions were made and elemental analysis was carried out using an ICPMS X series 2.

8.2.4.4 pH at the point of zero charge of the medium

The pH at the point of zero charge (pH_{pzc}) was determined by the potentiometric titration method (Appel et al., 2003). 5 mg of the crushed lava rock was added to 10 mL KNO_3 solution. 0.1 M HCl or 0.1 M NaOH was then added to 0.1, 0.01 and 0.001 M KNO_3 to have the pH of each electrolyte in the pH range of 3-11. The amounts of H^+ and OH^- adsorbed by a material were determined by subtracting the amount of HCl or

NaOH required to have the same pH for 10 mL of the electrolyte with no adsorbent added. The point of intersection of the titration curves (pH plotted against surface charge or amounts of acid or base added) for different concentrations of the electrolytes (varying ionic strength) was taken as pH_{pzc}.

8.2.4.5 Specific surface area of the crushed lava rock used

The specific surface area was determined based on the Brunaeur, Emmet and Teller (BET) theory with the Micromeritics ASAP® 2020 Accelerated Surface Area and Porosimetry Analyzer that uses the gas (nitrogen) sorption technique (Kalibbala et al., 2012; Sekomo et al., 2012). The volume of nitrogen gas (at a temperature of 77K) required to form a monolayer on the sample surface enabled the calculation of the surface area of a porous solid at a range of relative pressure values (p/p_o; the ratio of the pressure of the gas and pressure at saturation) of 0.05 - 0.3. The surface area occupied by a single molecule of adsorbate on the surface was calculated from the density of the liquefied adsorbate. The specific surface area was obtained for crushed lava rock of size class 2.56 - 5.00 mm and 1.18 - 2.56 mm used in the R1 and R2, respectively.

8.2.5 Sampling strategy

The grey water filter was operated for a period of five months from August 2012 to January 2013. Collection of samples was done in the morning at intervals of one day during the acclimatisation period (about 34 days) until a steady state was achieved based on COD and TOC removal. The samples for bacteriological tests were then collected for analysis twice a week, while samples for chemical parameters (COD, BOD_5, TP, Ortho-P, TKN, NH_4^+-N and NO_3^--N) were collected after every 2 days for analysis. After collection, the samples were stored in the dark at +4 °C using ice blocks and transported to the Public Health Engineering Laboratory at Makerere University (Uganda) for processing within one hour for bacteriological tests and within 12 hours for chemical parameters.

8.2.6 Analytical techniques

The pH, temperature, dissolved oxygen (DO) and electrical conductivity (EC) were measured using portable WTW (Wissenschaftlich-Technische Werkstätten) meters pH 3310; Oxi 3310 and Cond 3310. The total organic carbon (TOC) and dissolved organic carbon (DOC) of non-filtered and filtered samples, respectively, were determined using a TOC-5000A (Shimadzu, Milton Keynes, UK). Samples for measurement of DOC were filtered through 0.45μm cellulose nitrate membrane filters after leaching the filter DOC using 500 mL of distilled water in absence of water purified with resin filters and deionization. The total suspended solids (TSS) concentration was measured following standard methods (APHA, AWWA and WEF, 2005).

Chemical parameters (COD, BOD$_5$, TP, ortho-P, TKN, NH$_4^+$-N and NO$_3^-$-N) were determined using standard methods (APHA, AWWA and WEF, 2005). COD for non-filtered samples was determined using the closed reflux colorimetric method, BOD$_5$ using the HACH BOD track (serial no. 26197-01; Loveland, Co 80539, USA). The oil and grease concentration was obtained by solid phase extraction (SPE) using USEPA Method 1664A (USEPA, 1999). Kjeldahl Nitrogen (TKN) was determined using the Kjeldahl method, whereas NH$_4^+$-N and NO$_3^-$-N were determined using, respectively, the direct nesslerization method and the calorimetric method with Nitraver 5. Total phosphorus was determined using persulfate digestion followed by the ascorbic acid spectrophotometric method, while ortho-phosphorus was measured using the ascorbic acid spectrophotometric method. The numbers of *E. coli*, total coliforms and *Salmonella* spp. were determined with Chromocult® Coliform Agar media (Merck KGaA, Germany) using the spread plate method (APHA, AWWA and WEF, 2005).

8.3   Results

8.3.1   Physical and chemical characteristics of the crushed lava rock

The effective size (D$_{10}$) of the crushed lava rock in R1 and R2 was 1.25 mm and 2.85 mm, while the coefficient of uniformity (C$_u$) was 1.44 and 1.26, respectively (Table 8.1). The specific surface area (S$_{BET}$) of the crushed lava rock in R1 and R2 was 2.96 m^2.g^{-1} and 3.18 m^2.g^{-1}, respectively. There was no big difference in the porosity of the media in both filters, although the medium size in R1 was larger than that in R2 (Table 8.1). The crushed lava rock that was used as filter material was dominated by SiO$_2$ (46.8%), Al$_2$O$_3$ (15.4%), Fe$_2$O$_3$ (13.7%), CaO (11.1% and MgO (3.7%) based on X-ray diffraction analysis. The main elements in the crushed lava rock filter were Fe, Ca, Al, Mg and Na (Table 8.2). In addition, the crushed lava rock had a pH at zero charge (pH$_{zc}$) of about 7.1 (Table 8.1).

Table 8.1:   Characteristics of the crushed lava rock (filter medium)

Parameter	Unit	Parameter values of the crushed lava rock (filter media) used in this study		Reported parameter values of the crushed lava rock from same region	
		R1	R2	Sekomo et al. (2012)	Kalibbala et al. (2012)
Particle size range	mm	2.56 - 5	1.18 - 2.56	0.25 - 0.9	1.0 - 2.0
Effective size (D_{10})	mm	2.85	1.25	-	1.2
D_{30}	mm	3.45	1.65	-	-
D_{60}	mm	3.60	1.80	-	-
Coeficient of uniformity C_u	-	1.26	1.44	-	1.2
Cofficient of curvature C_c	-	1.16	1.21	-	-
Specific surface area, S_{BET}	$m^2 \cdot g^{-1}$	2.96	3.18	3.0	3.23
pH_{zc}	-	7.05	7.12	7.2	6.95
Bulk density	$kg.m^{-3}$	1,370	1,350	1,250	1,950
Particle density	$kg.m^{-3}$	3985	3560	3760	-
Porosity	%	65.6	62	66.8	-

Table 8.2: Elemental and chemical composition of the crushed lava rock (filter medium)

Element	Composition (mg/g)	Chemical compound	Composition (%)	Reported chemical composition of crushed lava rock (%)	
				Sekomo et al. (2012)	Alemayehu and Lennartz (2009)
Al	14.27 ±0.17	SiO_2	46.8	43.24	47.4
Ba	0.31±0.03	Al_2O_3	15.4	16.65	21.6
Ca	15.82±0.72	Fe_2O_3	13.7	14.15	8.9
Cu	0.86±0.02	CaO	11.1	10.95	12.4
Co	<0.05	MgO	3.7	4.29	3.3
Fe	36.52±1.35	TiO_2	2.9	3.57	1.7
K	6.52±0.66	Na_2O	2.3	2.44	3.0
Mg	13.25±0.19	K_2O	2.0	2.35	0.5
Mn	0.645±0.03	P_2O_5	-	0.77	-
Na	3.59±0.1	BaO	0.3	0.40	-
Ni	0.09±0.01	MnO	0.3	0.23	-
Zn	0.05±0.00				

8.3.2   Characteristics of raw and pre-treated grey water

The high standard deviation of the concentration values of the chemical and bacteriological parameters of grey water (Table 8.3) shows the variability of the quality of both non-settled and settled grey water. Sedimentation of the grey water generated at the household reduced the COD concentration by 51%; the BOD_5 concentration by 26.5% and the TSS concentration by 59.5%. The oil and grease concentration reduced by 74% after skimming off the top layer from the grey water, while there was no major change in the concentration of the chemical parameters and the bacteriological quality of grey water as a result of sedimentation (Table 8.3).

Table 8.3: Physical and chemical characteristics of the non-settled and settled grey water that was treated with a crushed lava rock filter

Parameter	unit	Parameter values		Reported parameter values of grey water in peri-urban areas	
		Non-settled grey water (n=30)	Settled grey water (n=30)	Sall and Takahashi (2006)	Kulabako et al. (2012)
Temperature	°C	23.5±2.2	24.0±1.5	22 - 26.5	21 - 28
pH	-	7.6±1.5	7.1±2.3	6.7 - 7.2	-
EC	$\mu S.cm^{-1}$	2150±147	1979±281	110 - 4100	500 - 1200
COD	$mg.L^{-1}$	6563±1864	3194±785	114.9 - 3229	1200 - 3400
BOD_5	$mg.L^{-1}$	1395±466	1025±625	115 - 1225	250 - 800
TOC	$mg.L^{-1}$	940±161	892±124	20.5 - 858	-
DOC	$mg.L^{-1}$	568±102	559±121	-	-
TSS	$mg.L^{-1}$	2828±735	1145±437	741 - 3180	-
Oil and grease	$mg.L^{-1}$	26.0±7.1	6.7±2.1	-	-
NH_4^+-N	$mg.L^{-1}$	29.5±7.9	26.7±6.3	47 - 110	5.1 - 12
NO_3^--N	$mg.L^{-1}$	2.7±0.5	3.3±0.3	1.5 - 5.5	-
TKN	$mg.L^{-1}$	34.8±9.2	29.5±4.7	-	-
Total-P	$mg.L^{-1}$	6.2±0.6	5.8±0.3	36.3 - 136	1.5 - 7.3
Ortho-P	$mg.L^{-1}$	3.9±1.4	3.4±1.5	11.8 - 44	0.5-3.5
Total coliforms	$cfu.(100\ ml)^{-1}$	$8.72 \times 10^7 (\pm 1.8 \times 10^7)$	$8.98 \times 10^7 (\pm 1.5 \times 10^7)$	$3.1 \times 10^7 - 2.1 \times 10^8$	-
Escherichia coli	$cfu.(100\ ml)^{-1}$	$3.70 \times 10^6 (\pm 3.8 \times 10^6)$	$3.20 \times 10^6 (\pm 2.9 \times 10^6)$	$4.1 \times 10^6 - 1.5 \times 10^7$	$0 - 1.4 \times 10^4$
Salmonella spp.	$cfu.(100\ ml)^{-1}$	$2.73 \times 10^4 (\pm 1.80 \times 10^4)$	$2.80 \times 10^4 (\pm 1.3 \times 10^4)$	-	-

8.3.3   Pollutant removal from grey water by the crushed lava rock filter

8.3.3.1   COD, TOC, DOC and TSS removal

The filter unit was operated for 41 days to allow for steady state conditions to be achieved during the filter usage. A stable COD and DOC removal efficiency of, respectively, 90% and 70% were achieved after 35 days of filter operation (Figure 8.3). The COD and TSS removal efficiency during uncontrolled conditions (HLR = 0.5 to 1.1 $m.d^{-1}$) after the ripening period was 85.9% and 87.5%, respectively (Figure 4A). The organic loading rate varied from 41 g $BOD_5.m^{-2}.d^{-1}$ to 115 g $BOD_5.m^{-2}.d^{-1}$. The COD removal efficiencies by R1 and R2 were 70% and 69%, while the TSS removal efficiencies were 82% and 78%, respectively (Figure 8.4A). When the HLR was made constant and reduced to 0.39 $m.d^{-1}$, the COD and TSS removal efficiencies increased to 90.5% and 94%, respectively (Figure 8.4B). The COD and TSS concentration of the effluent was, respectively, 256 ± 42 $mg.L^{-1}$ and 68 ± 17 $mg.L^{-1}$ at a HLR of 0.39 $m.d^{-1}$. This represents a high reduction in pollution load from the organic material when these values are compared to the COD and TSS concentration of the influent (settled grey water) of 3194±785 $mg.L^{-1}$ and 1145±437 $mg.L^{-1}$, respectively.

The TOC and DOC removal efficiencies were 68% and 66% under uncontrolled conditions (HLR = 0.5 to 1.1 $m.d^{-1}$) and 69.5% and 66.7% under controlled conditions (HLR= 0.39 $m.d^{-1}$), respectively. The highest TOC removal efficiencies of R1 and R2 were 45% and 52%, while the highest DOC removal efficiencies were 41% and 48%, respectively, at a HLR of 0.39 $m.d^{-1}$. During the filter operation, the initial filtration rate of 1.1 $L.min^{-1}$ decreased to 0.5 $L.min^{-1}$ after 30 days of filter operation when the filter unit was loaded with grey water. When the filtration rate was reduced to about 0.4 $L.min^{-1}$, scarification of the top of the media was done and the infiltration rate increased back to about 0.8 $L.min^{-1}$ and decreased again with time (Figure 8.5).

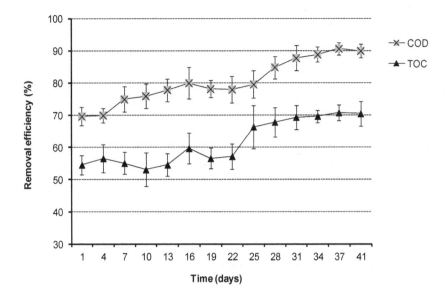

Figure 8.3: Evolution of the COD and TOC removal efficiency during the ripening period (Hydraulic loading rate: 0.5 to 1.1 m.d^{-1})

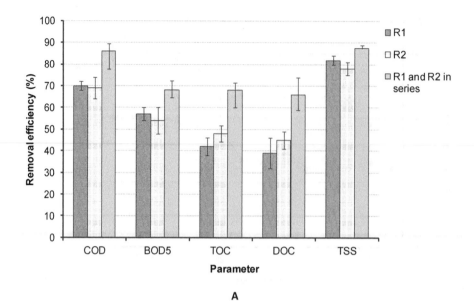

A

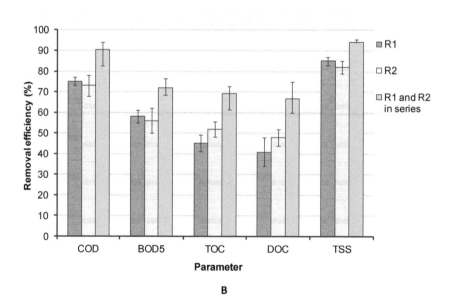

B

Figure 8.4:   Removal efficiency of COD, BOD$_5$, TOC, DOC and TSS during grey water treatment by a two-step crushed lava rock filter after the ripening period at (A) a hydraulic loading rate (HLR) varying from 0.5 to 1.1 m.d^{-1} and (B) a constant HLR of 0.39 m.d^{-1}

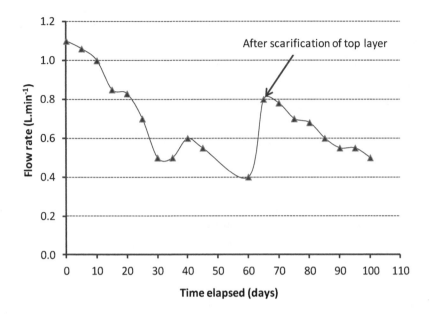

Figure 8.5: The flow rate at the outlet of reactor 2 after grey water charge during operation of the grey water filter after the ripening period (Hydraulic loading rate: 0.5 to 1.1 m.d^{-1})

8.3.3.2 Nutrient removal

The crushed lava rock filter unit was monitored for removal of TKN, NH$_4^+$-N, NO$_3^-$-N, TP and Ortho-P. The removal efficiency of TKN by R1 and R2 at varying HLR (from 0.5 to 1.1 m.d^{-1}) was 65.5%, while at a constant HLR (0.39 m.d^{-1}) the removal efficiency was 69% (Figure 8.6). The highest TKN removal efficiency by R1 and R2 was 58% and 62% , respectively, at a constant HLR (0.39 m.d^{-1}). The filter influent was 29.7 ± 4.7 mg.L^{-1} (Table 8.3), while the effluent TKN concentration after ripening period at a varying HLR and a constant HLR was 10.3 ± 1.4 mg.L^{-1} and 9.1 ± 1.3 mg.L-1, respectively. The highest removal efficiencies of NH$_4^+$-N and NO$_3^-$-N were 68% and 87% at a constant HLR. There was a higher removal of NO$_3^-$-N when compared to the other nutrient components; the influent and effluent concentrations were 3.3 ± 0.3 mg.L^{-1} and 0.29 ± 0.04 mg.L^{-1}. Operating the filter unit at a constant HLR (0.39 m.d^{-1}) from a varying HLR (0.5 to 1.1 m.d^{-1}), led to a minor removal efficiency increase of TP and ortho-P (from 58% to 59.5% and 53% to 54%, respectively; Figure 8.6). The corresponding TP removal efficiencies by R1 and R2 were 39% and 44%, respectively. The impact of the reduced and constant HLR was lower on the nutrient removal than on the organic matter removal.

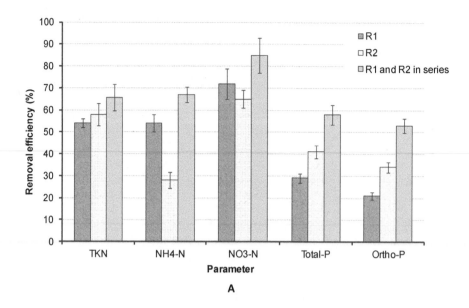

A

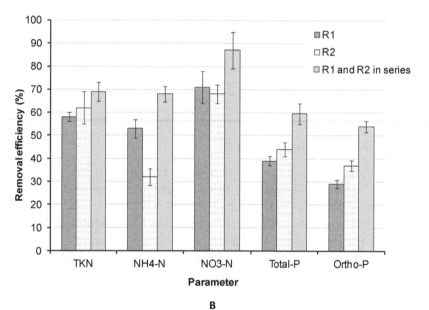

B

Figure 8.6: Removal efficiency of nutrients during grey water treatment by a two-step crushed lava rock filter after the ripening period at (A) a hydraulic loading rate (HLR) varying from 0.5 to 1.1 m.d^{-1} and (B) a constant HLR of 0.39 m.d^{-1}

8.3.3.3 Microorganisms' removal

The removal of *E. coli, Salmonella* spp. and total coliforms was determined during the operation of the crushed lava rock filter unit. The log removal of *E. coli* by R1 and R2 were 2.21 and 1.59, respectively, while the resultant log removal by R1 and R2 in series was 3.81 at a HLR of 0.5 to 1.1 m.d^{-1} based on R1 influent and R2 effluent *E. coli* concentrations (Table 8.4). The log removal of *Salmonella* spp. R1 and R2 was 1.90 and 1.30, respectively. In addition, the log removal of *Salmonella* spp. and total coliforms by R1 and R2 was 3.20 and 3.92, respectively. At a HLR of 0.39 m.d^{-1}, the log removal of *E. coli*, *Salmonella* spp. and total coliforms by R1 and R2 in series was 3.9, 3.5 and 3.9, respectively. Despite the log removal greater than 3, the concentration of *E. coli, Salmonella* spp. and total coliforms in the R2 effluent was 2.95 x 10^3 cfu.(100 mL)$^{-1}$, 5.30 x 10^2 cfu.(100 mL)$^{-1}$ and 3.38 x 10^4 cfu.(100 mL)$^{-1}$, respectively (Table 8.4).

Table 8.4: Log removal of *E. coli*, *Salmonella* spp. and total coliforms after ripening period during grey water treatment with a crushed lava rock filter unit installed at household level

Sample	Treatment process	Bacteriological quality[†]			Log removal[††]		
		E. coli (n=27)	*Salmonella* spp. (n=27)	TC (n=27)	*E. coli*	*Salmonella* spp.	TC
Non-settled grey water (SP1)		1.91E+07 (±1.63E+07)	8.33E+05 (±1.89E+06)	2.79E+08 (±2.28E+08)			
Effluent from SP2	Sedimentation followed by filtration through R1[*]	1.16E+05 (±1.37E+05)	1.06E+04 (±1.45E+04)	3.21E+06 (±2.81E+06)	2.218	1.896	1.939
Effluent from SP3	Filtration through R2[**]	2.95E+03 (±2.72E+03)	5.2E+02 (±3.47E+02)	3.38E+04 (±1.62E+04)	1.595	1.300	1.976
	Filtration through R1 and R2[***]				3.813	3.196	3.916

[†] Concentrations are in cfu/100 ml and in brackets is the standard deviation. TC refers to total coliforms

[††] Log removal at HLR of 0.5 to 1.1 m.d-1

[*] Coarse filter: media of crushed lava rock of depth 30 cm and size 2.36 - 5 mm supported by 10 cm depth of gravel.

[**] Second filter in series with coarse filter: media of crushed lava rock of depth 30 cm and size 1.18- 2.36 mm supported by 10 cm depth of gravel.

[***] Log removal of *E. coli*, *Salmonella* spp. and total coliforms at a constant hydraulic loading rate of 0.39 m.d-1 was 3.9, 3.5 and 3.9, respectively.

220

8.4   Discussion

8.4.1   Use of the crushed lava rock filter for grey water treatment in slums

This study demonstrated that there is potential for application of low-technology grey water treatment at household level in urban slums. The pollutant (organic matter, nutrients and bacteria) removal efficiencies achieved by treating grey water with a crushed lava rock filter (Figures 8.4 and 8.6; Table 8.4) indicate that wide application of such filters at household level in slums can significantly reduce the pollutant loads from the grey water stream. The active participation of the household members is critical to have high grey water collection efficiency and to avoid spillage of grey water or misuse of the filter unit, which may reduce the filter performance. Moreover, oil and grease and solid particles cause filter clogging when grey water is not pre-treated resulting in bad odours.

The crushed lava rock filter was first monitored until a steady sate was reached. The COD and TOC removal efficiencies increased with operation time of the filter until after 35 days when maximum removal efficiencies were achieved (Figure 8.3). Pollutant removal processes in media based systems increase with the filter age as a result of growth of biofilms on the filter media, straining, particle agglomeration, and sedimentation during the acclimatisation period (Campos et al., 2002; Crites and Tchobanoglous, 1998; Jellison et al., 2000). The ripening period of the crushed lava rock was within the reported ripening period range of 5 - 8 weeks for wastewater treatment with media (gravel, sand) filters (Achak et al., 2009; Joubert, 2008; Rodgers et al., 2005).

The crushed lava rock filter was first operated at a hydraulic loading rate varying from 0.5 to 1.1 $m.d^{-1}$ and then at a constant hydraulic loading rate of 0.39 $m.d^{-1}$. There was an increase in the COD, BOD_5 and TSS removal of 5%, 4% and 7% when the filter was operated at a constant HLR of 0.39 $m.d^{-1}$, while there was no significant difference in the removal of nutrients ($p = 0.34 >$ significant level $p = 0.05$). An increase in both hydraulic loading rate and organic loading rate have been found to reduce the effluent quality in media based systems (Essandoh et al., 2011; Achak et al., 2009; Healy et al., 2007). In this study, the effect of the varying organic loading rate as a result of the variable grey water quality was reduced by allowing the grey water to settle for an hour prior to feeding it to the crushed lava rock filter unit. In addition, the use of a two-step filtration process using R1 and R2 with coarse (2.56-5 mm) and fine (1.18-2.56 mm) medium also led to less impact of organic loading rate on the effluent quality.

The cost of the crushed lava rock filter unit of US$ 250 (based on rates of December 2012) is not affordable by the urban poor in slums. The total cost of the filter unit can be reduced by use of cheaper materials depending on the income level of a household. The use of brick work for filter support, plastic buckets that are common in slum households and limiting the number of pipe fittings are attractive options. An extra source of income could be established through reuse: the effluent could be further treated in a tertiary step of a grey water tower garden on which vegetables are grown for consumption (Kulabako et al., 2011; Morel and Diener, 2006). The solids after sedimentation may be dewatered and dried under the sun (feasible in tropics) for pathogen removal prior to use as a soil conditioner. Yet, these options may not recover all costs.

Alternatively, the filter effluent may be used for indirect non-potable uses such as infiltration in the gardens (during subsurface irrigation) or for preparation of mortar during construction. Treated grey water may be discharged in the tertiary in case of no reuse. The reduction of the grey water pollutant loads discharged into the environment by the low-technology crushed lava rock filter results in reduced risks to public health. Moreover, improved sanitary conditions lead to improved health status and quality of life (Isunju et al., 2011; Okun, 1998). Decentralised treatment of grey water such low-technology solutions that occupy less space (< 1.5 m^2) is the only option because it is difficult to have a centralised wastewater treatment in the slum with limited space and where the ability to pay for services is low.

8.4.2   Performance of the crushed lava rock filter

The low biodegradability (BOD$_5$: COD = 0.32; Table 8.3) of grey water generated in Bwaise III, indicated that physical straining was a key removal mechanism for organic matter during filtration. The straining process removes particles in grey water with a size larger than the pore space. The COD and TSS removal efficiency under both controlled and uncontrolled conditions (Figures 8.4A and 8.4B) were within the reported range of 70 - 95% during grey water treatment with sand and gravel media filters (Al-Hamaiedeh and Bino, 2010; Dalahmeh et al., 2012; Mandal et al., 2011). The removal of biodegradable organic matter by 68% to 72% (Figure 8.4) is attributed to the biomass activity on the filter surface. The BOD$_5$ removal obtained in this study was lower than the reported range of 75 - 92% (Dalahmeh et al., 2012; Rodgers et al., 2005). This was attributed to the differences in the level of biodegradability of the grey water (Katukiza et al., 2013a) and the variability of the influent hydraulic loading rate (which depends on household generation pattern) to the crushed lava rock filter.

The pollutant removal efficiencies by R1 and R2 under varying HLR and a constant HLR (Figures 8.4 and 8.6), was in all cases higher than that of either R1 or R2 and thus having a two-step filtration process results in a better effluent quality than using one filter. The removal of dissolved Kjeldahl nitrogen and phosphorus was mainly attributed to precipitation. Adsorption and ionic exchange have been found to contribute to the removal of phosphates from wastewater (Mann and Bavor, 1993). The removal of dissolved Kjeldahl nitrogen and phosphorus by adsorption was limited since the pH of grey water was higher than the pH$_{zc}$ (7.05) of the crushed lava rock. In addition, particulate Kjeldahl nitrogen was probably removed by straining during the filter operation. The results show that NH_4^+-N removal was probably due to adsorption (Achak et al., 2009). The NH_4^+-N and NO_3^--N concentration amounting to 22.8 ± 4.3 mg.L^{-1}, and 0.3 ± 0.04 mg.L^{-1}, respectively, in the R1 effluent indicate that there was limited nitrification and the nitrate in the influent was denitrified. Nutrient removal by adsorption on the crushed lava rock was probably limited because the pH of the grey water was above 7.1 (pH$_{zc}$). Ammonia volatilisation was most unlikely because the pH of R1 and R2 effluents varied between 6.2 to 6.6, yet NH_4^+ is transformed into volatile NH_3 at a higher pH (above 8) (Hammer and Knight, 1994).

The removal of microorganisms from the grey water by the crushed lava rock filter was due to a combination of factors including biological activity (predation, die-off), exposure to sunlight, UV radiation, and retention on the filter medium by straining and adsorption. Auset et al. (2005) also found that colloids attachment to the solid water interface was responsible for high removal of microorganisms in a porous media during intermittent filtration of primary and secondary municipal effluents. In addition, protozoa grazing may be an important bacterial removal mechanism (Bomo et al., 2004). The log removal range of 1.3 to 2.2 by either R1 or R2 and the > 3 Log removal by R1 and R2 of *E. coli, Salmonella* spp. and total coliforms, respectively (Table 8.4), are comparable to the reported log removal range of 1 to >3 for bacteria in media (sand soil, mulch) filters (Elliott et al., 2008; Hijnen et al., 2004). The results obtained in this study show that a higher log removal of *E. coli, Salmonella* spp. and total coliforms was achieved by a series of R1 and R2 than by either R1 or R2 alone (Table 8.4).

There is a need for further investigation on the performance of the crushed lava rock filter regarding the removal of viruses. Grey water contains viruses (Katukiza et al., 2013), which may pose a risk to users of the grey water filter unit because of their ability to adaptation to different environmental conditions (Baggi et al., 2001; Fong et al., 2010). The removal of protozoa and helminth eggs was not investigated in this study. Previous studies have found that they are effectively removed by media (sand, soil, lava) filters (Bomo et al., 2004; Hijnen et al., 2004). The presence of micro-pollutants in grey is another issue that requires further investigation. Micro-pollutants originate from household chemicals, pharmaceuticals and hormones originating from urine contamination (Hernández-Leal et al., 2011; Larsen et al., 2009).

8.4.3   Characteristics of the filter medium (crushed lava rock)

The coefficient of uniformity values for the crushed lava rock meet the criterion of $C_u<$ 4 required for adequate hydraulic conductivity to reduce the risk of clogging (Craig, 1997). The medium was also well graded since the coefficient of curvature (C_c) was in the recommended range of 1 to 3 (Healy et al., 2007). The specific surface area (about 3 m2.g-1; Table 8.1) of the crushed lava rock used in this study was comparable to values obtained in previous studies (Table 8.1). In addition, the chemical composition of the crushed lava rock was comparable to properties of crushed lava rock obtained in previous studies (Table 8.2). The presence of Fe, Ca, Al, Mg and their compounds in crushed lava rock suggests the occurrence of adsorption and ionic exchange as pollutant removal mechanisms during the filter operation. The pH at zero charge (pH_{zc}) of the crushed lava rock of about 7.1 (Table 8.1) and the pH of the grey water exceeding 7.1 (Table 8.3) show that the surface of the crushed lava rock attracted cations like NH_4^+ present in the grey water, while anions like NO_3^- and PO_4^{3-} were repelled. The determined media characteristics therefore influenced the pollutant removal mechanisms during grey water treatment using a crushed lava rock filter.

8.4.4   Effect of clogging on the filtration rate

Clogging of the crushed lava rock filter was on one hand attributed to the accumulation of particulate organic matter and oil and grease present in the grey water, and biofilm formation on the media surface on the other hand. These reduce the filter permeability (Cammarota and Freire, 2006; Leverenz et al., 2009; Rodgers et al., 2004). Variability in concentration of organic substrate and biomass development to support the biofilm are other factors that contribute to clogging (Campos et al., 2002; Elliott et al., 2008). Clogging results in head-loss development and reduction of the filtration rate as a result of reduced pore space (Healy et al., 2007). Increasing the influent COD concentration and filter dosing frequency and hydraulic loading rate were found to reduce the filter performance due to clogging (Leverenz et al., 2009)

In this study, filter clogging effects were minimised by the sedimentation of raw grey water as indicated by the reduction of TSS, and oil and grease concentration (Table 8.3). Whenever clogging occurred during the filter operation and the R2 effluent flow reduced to 0.4 $L.min^{-1}$, scarification of the top 5cm of the filter bed in reactor 1 was done to overcome the problem of low filtration rate (Figure 8.5). In addition, the wetting and drying cycles of the filter surface as a result of intermittent flow helped to increase the filtration rate and thus enable long operational periods (4-6 weeks) at a flow rate of at least 0.5 $L.min^{-1}$. However, the initial filtration rate of 1.1 $L.min^{-1}$ could not be restored even at a hydraulic loading rate of 0.5 $m.d^{-1}$. Filters operating at low dosing frequencies have been found to sustain high treatment efficiencies with limited maintenance (Leverenz et al., 2009) and thus intermittent flow during the filter operation could have increased the filter oxidation capacity and minimised clogging effects.

8.5   Conclusions

- The implementation and use of a crushed lava rock filter unit for grey water treatment at household level in Bwaise III slum, demonstrated that low-technology onsite grey water treatment has the potential to reduce the grey water pollutant loads in urban slums. This may complement excreta management interventions to achieve adequate sanitation in urban slums and thus improve the quality of life of a slum dweller.

- The hydraulic loading rate was identified as a key design parameter for the crushed lava rock filter. The crushed lava rock filter achieved COD and TSS removal efficiencies of 85-88% at varying HLR of 0.5 to 1.1 m.d^{-1} and 90-94 % at a constant HLR of 0.39 m.d^{-1}. In addition, the highest removal efficiencies of TP and TKN of 59.5% and 69% were achieved at a HLR of 0.39 m.d^{-1}. The organic loading rate also impacts the effluent quality although this was minimised in this study by sedimentation of the grey water and the use of a two-step filtration process.
- Using a two-step filtration process by operating filters R1 and R2 in series resulted in higher pollutant removal efficiencies than using a single filtration step with R1 only.
- A log removal > 3 (99.9%) of *E. coli*, *Salmonella* spp. and total coliforms was achieved under household filter usage conditions. However, the concentration of these microorganisms in the final effluent did not meet the WHO guidelines for unrestricted wastewater reuse in agriculture and aquaculture. There is a need to investigate the feasibility of adding a tertiary treatment step to increase the removal of these microorganisms.
- The filtration rate reduced to less than 50% of its original value after 30 days of filter operation. The low dosing frequency as a result of intermittent flow conditions and the pre-treatment of grey water to remove TSS and oil and grease sustained the filter performance with limited filter maintenance.
- Further studies on the removal of viruses and micro-pollutants by the crushed lava rock filter are required.

References

Abu Ghunmi, L., Zeeman, G., van Lier, J., Fayyed, M., 2008. Quantitative and qualitative characteristics of grey water for reuse requirements and treatment alternatives: the case of Jordan. Water Science and Technology 58(7), 1385-1396.

Achak, M., Mandi, L., Ouazzani, N., 2009. Removal of organic pollutants and nutrients from olive mill wastewater by a sand filter. Journal of Environmental Management 90(8), 2771-2779

Alemayehu, E., Lennartz, B., 2009. Virgin volcanic rocks: kinetics and equilibrium studies for the adsorption of cadmium from water. Journal of Hazardous Materials 169, 395-401.

Al-Hamaiedeh, H., M. Bino, M., 2010. Effect of treated grey water reuse in irrigation on soil and plants. Desalination 256, 115-119

APHA, AWWA and WEF., 2005. Standard Methods for the Examination of Water and Wastewater. *American Public Health Association, American Water Works*

Association and Water Environment Federation publication, 21st edition. Washington DC, USA.

Appel, C., Ma, L. Q., Rhue, R. D., Kennelley, E., 2003. Point of zero charge determination in soils and minerals via traditional methods and detection of electroacoustic mobility. Geoderma 113(1-2), 77–93.

ASTM., 1998. Manual on test sieving methods: guidelines for establishing sieve analysis procedures. Lawrence R. P., Charles, W. W. (Ed.), West Conshohocken: American Society for Testing and Materials.

Auset, M., Keller, A.A., Brissaud, F., Lazarova, V., 2005. Intermittent filtration of bacteria and colloids in porous media. Water Resources Research 41(9) W09408:1-13

Baggi, F., Demarta, A., Peduzzi, R., 2001. Persistence of viral pathogens and bacteriophages during sewage treatment: lack of correlation with indicator bacteria. Research in Microbiology 152, 743–751.

Birks, R., Hills, S., 2007. Characterisation of indicator organisms and pathogens   in domestic greywater for recycling. Environmental Monitoring and Assessment 129, 61–69.

Bomo, A.M, Stevik, T.K, Hovi, I., Hanssen, J.F., 2004. Bacterial removal and protozoan grazing in biological sand filters. Journal of Environmental Qualty 33(3), 1041-7.

Cammarota, M.C., Freire, D.M.G., 2006. A review on hydrolytic enzymes in the treatment of wastewater with high oil and grease content. Bioresource Technology 97, 2195-2210.

Campos, L.C., Su, M.F.J., Graham, N.J.D., Smith, S.R., 2002. Biomass development in slow sand filters. Water Research 36(18), 4543-4551.

Carden, K., Armitage, N., Sichone O., Winter, K., 2007b. The use and disposal of grey water in the non-sewered areas of South Africa: Part 2- Grey water management options. Water SA 33 (4), 433-442.

Carden, K., Armitage, N., Winter, K., Sichone, O., Rivett, U., Kahonde, J., 2007a. The use and disposal of grey water in the non-sewered areas of South Africa: Part 1- Quantifying the grey water generated and assessing its quality. Water SA 33(4), 425-432.

Craig, R.F., 1997. Soil Mechanics, seventh ed. E&FN Spon, London, 447pp.

Crites, R.W., 1998. Tchobanoglous, G. Small and Decentralized Wastewater Management Systems. McGraw-Hill Companies, California, ISBN 0-07-289087-8 USA.

Dalahmeh, S.S., Lars D. Hylander, L.D., Vinnerås, B., Pell, M., Öborn, I., Jönsson, H., 2011. Potential of organic filter materials for treating greywater to achieve irrigation quality: a review. Water Science and Technology 63(9), 2011.

Dalahmeh, S.S., Pell, M., Vinnerås, B., Hylander, L.D., Öborn, I., Håkan Jönsson, H, 2012. Efficiency of bark, activated charcoal, foam and sand filters in reducing pollutants from greywater. Water Air and Soil Pollution 223(7), 3657–3671

Elliott M.A., Stauber, C.E., Koksal, F., DiGiano, F.A., Sobsey, M.D., 2008. Reductions of *E. coli*, echovirus type 12 and bacteriophages in an intermittently operated household-scale slow sand filter. Water research 42(10-11), 2662–2670.

Eriksson, E., Andersen, H.R., Madsen, T.S., Ledin, A., 2009. Grey water pollution variability and loadings. Ecological Engineering 35(5), 661-669.

Escher, B.I., and Fenner, K., 2011. Recent Advances in Environmental Risk Assessment of Transformation Products. Environmental Science and Technology 45, 3835-3847.

Fong, T.T, Phanikumar, M.S., Xagoraraki, I., Joan B. Rose, J.B., 2010. Quantitative Detection of Human Adenoviruses in Wastewater and Combined Sewer Overflows Influencing a Michigan River. Applied and Environmental Microbiology 76, 715–723.

Gross, A., Azulai, N., Oron, G., Ronen, Z., Arnold, M., Nejidat, A., 2005. Environmental impact and health risks associated with greywater irrigation: a case study. Water Science & Technology 52(8), 161–169.

Gross, A., Shmueli, O., Ronen, Z., Raveh, E., 2007. Recycled vertical flow constructed wetland (RVFCW) - a novel method of recycling greywater for irrigation in small communities. Chemosphere 66(5), 916–23.

Hammer, D.A., Knight, R.L., 1994. Designing constructed wetlands for nitrogen removal. Water Science Technology 29(4), 15–27.

Healy, M.G., Rodgers, M., Mulqueen, J., 2007. Treatment of dairy wastewater using constructed wetlands and intermittent sand filters. Bioresource Technology 98(12), 2268-2281.

Hernández-Leal, L., Temmink, H., Zeeman, G., Buisman, C.J.N., 2011. Removal of micropollutants from aerobically treated grey water via ozone and activated carbon. Water Research 45(9), 2887-2896.

Hijnen, W.A, Schijven. J.F, Bonné, P., Visser, A., Medema, .G.J., 2004. Elimination of viruses, bacteria and protozoan oocysts by slow sand filtration. Water Science and Technology 50(1), 147-54.

Isunju, J.B, Etajak, S., Mwalwega, B., Kimwaga, R., Atekyereza, P., Bazeyo, W., Ssempebwa, J.C., 2013. Financing of sanitation services in the slums of Kampala and Dar es Salaam. Health 5(4), 783-791.

Isunju, J.B., Etajak, S., Mwalwega, B., Kimwaga, R., Atekyereza, P., Bazeyo, W., John C. Ssempebwa, J.C., 2013. Financing of sanitation services in the slums of Kampala and Dar es Salaam. Health 5 (4), 783-791.

Jamrah, A., Al-Futaisi, A., Prathapar, S., Harrasi, A.A., 2008. Evaluating greywater reuse potential for sustainable water resources management in Oman. Environmental Monitoring and Assessment 137(1-3), 315-327

Jefferson, B., Burgess, J.E., Pichon, A., Harkness, J., Judd, S.J., 2001. Nutrient addition to enhance biological treatment of greywater. Water Research 35 (11), 2702–2710.

Jellison, K.L., Richard I. Dick, R.I., Monroe L. Weber-Shirk, M.L., 2000.. Enhanced ripening of slow sand filters. Journal of Environmental Engineering 126(12), 153-1157

Joubert, E.D., 2008. Visualisation of the microbial colonisation of a slow sand filter using an Environmental Scanning Electron Microscope. Electronic Journal of Biotechnology 11(2), 1-7.

Kalibbala, H.M., Wahlberg, O., Plaza, E., 2012. Horizontal flow filtration: impact on removal of natural organic matter and iron co-existing in water source. Separation Science and Technology 47, 1628-1637.

Katukiza, A.Y, Ronteltap, M., Niwagaba, C., Kansiime, F., Lens, P.N.L., 2010. Selection of sustainable sanitation technologies for urban slums - A case of Bwaise III in Kampala, Uganda, Science of the Total Environment 409(1), 52-62.

Katukiza, A.Y, Ronteltap, M., Niwagaba, C., Kansiime, F., Lens, P.N.L., 2013a. Grey water characterisation and pollution load in an urban slum. International Journal of Environmental Science and Technology."Accepted".

Katukiza, A.Y., Temanu H, Chung JW, Foppen JWA, Lens PNL., 2013b. Genomic copy concentrations of selected waterborne viruses in a slum environment in Kampala, Uganda. Journal of Water and Health 11(2), 358-369.

Kulabako, N. R., Ssonko, N.K.M, Kinobe, J., 2011. Greywater Characteristics and Reuse in Tower Gardens in Peri-Urban Areas – Experiences of Kawaala, Kampala, Uganda. The Open Environmental Engineering Journal 4, 147-154.

Larsen, T.A., Alder, A.C., Eggen, R.I.L., Maurer, M., Lienert, J., 2009. Source separation: will we see a paradigm shift in wastewater handling? Environmental Science and Technology 43(16), 6121-6125.

Leverenz, H.L., Tchobanoglous, G., Darby, J.L., 2009. Clogging in intermittently dosed sand filters used for wastewater treatment. Water Research 43(3), 695-705.

Li F, Wichmann K, Otterpohl R., 2009. Review of technological approaches for grey water treatment and reuses. Science of the Total Environment 407(11), 3439-3449.

Mandal, D., Labhasetwar, P., Dhone, S., Dubey, A.S., Shinde, G., Satish Wate, S., 2011. Water conservation due to greywater treatment and reuse in urban setting with specific context to developing countries. Resources, Conservation and Recycling 55, 356–361.

Mann, R.A., Bavor, H.J., 1993. Phosphorous removal in constructed wetlands using gravel and industrial waste substance. Water Science and Technology 27,

Masi, F., El Hamouri, B., Abdel Shafi, H., Baban, A., Ghrabi, A., Regelsberger, M., 2010. Treatment of segregated black/grey domestic wastewater using constructed wetlands in the Mediterranean basin: the zer0-m experience. Water Science and Technology 61(1), 97-105

Metcalf, Eddy., 2003. Wastewater engineering: treatment and reuse. 4th ed. 1221 Avenue of Americas, New York, NY 10020, USA: McGraw-Hill Companies.

Morel, A., Diener, S., 2006. Greywater Management in Low and Middle-Income Countries. Review of different treatment systems for households or neighbourhoods.
http://www.eawag.ch/forschung/sandec/publikationen/ewm/dl/GW_managem ent.pdf [Accessed on 3rd January, 2013].

Nyenje, P.M., Foppen, J.W., Uhlenbrook, S., Kulabako, R., Muwanga, A., 2010. Eutrophication and nutrient release in urban areas of sub-Saharan Africa: a review. Science of the Total Environment 408(3), 447-455.

Ottoson, J., Stenström, T.A., 2003. Faecal contamination of grey water and associated microbial risks. Water Research 37(3), 645-655.

Panuvatvanich, A., Koottatep, T., Kone, D., 2009. Influence of sand layer depth and percolate impounding regime on nitrogen transformation in vertical-flow constructed wetlands treating faecal sludge. Water Research 43, 2623-2630.

Panuvatvanich, A., Koottatep, T., Kone, D., 2009. Influence of sand layer depth and percolate impounding regime on nitrogen transformation in vertical-flow constructed wetlands treating faecal sludge. Water Research 43, 2623-2630.

Rodgers, M., Healy, M.G., Mulqueen, J. 2005. Organic carbon removal and nitrification of high strength wastewaters using stratified sand filters. Water Research 39(14), 3279–3286

Rodgers, M., Mulqueen, J., Healy, M.G., 2004. Surface clogging in an intermittent stratified sand filter. Soil Science Society of America Journal 68, 1827-1832.

Rühlmanna, J., Körschens, M., Graefe, J., 2006. A new approach to calculate the particle density of soils considering properties of the soil organic matter and the mineral matrix. Geoderma, 130 (3-4), 272–283.

Sall, O., Takahashi, Y.. 2006. Physical, chemical and biological characteristics of stored greywater from unsewered suburban Dakar in Senegal. Urban Water Journal 3(3), 153-164.

Schwarzenbach, R.P., Egli, T., Hofstetter, T.B., von Gunten, U., Wehrli, B., 2010. Global Water Pollution and Human Health. Annual Review of Environment and Resources 35, 109–36

Sedlak, R., 1991. Phosphorus and Nitrogen Removal from Municipal Wastewater: Principles and Practice. 2nd ed. Lewis Publishers, New York, USA.

Sekomo, C.B., Rousseau, D.P. L., Lens, P.N.L., 2012. Use of Gisenyi Volcanic Rock for Adsorptive Removal of Cd(II), Cu(II), Pb(II), and Zn(II) from. Water Air and Soil Pollution, 223, 533–547

Ternes, T., Joss, A., 2006. Human pharmaceuticals, hormones and fragrances: the challenge of micropollutants in urban water management. IWA Publishing, London.

Travis, M.J., Weisbrod, N., Gross, A., 2008. Accumulation of oil and grease in soils irrigated with greywater and their potential role in soil water repellency. Science of the Total Environment 394, 68-74.

USEPA., 2009. Method 1664 revision A: N-hexane extractable material (HEM; oil and grease) and silica gel treated N-hexane extractable material (SGT-HEM; non-polar material) by extraction and gravimetry. Washington DC: United States Environmental Protection Agency.

Von Felde, K., Kunst, S., 1997. N- and COD-removal in vertical-flow systems. Water Science and Technology 35 (5), 79–85.

von Sperling, M. and Chernicharo, C.A.L., 2005. Biological Wastewater Treatment in Warm Climate Regions, IWA Publishing, London, 835 pp.

Chapter 9:   General discussion, conclusions and recommendations

9.1   Introduction

Environmental pollution and public health risks in urban slums due to inadequate and poor sanitation is one of the major challenges to urban authorities in developing countries. The limited sanitation interventions in urban slums have mainly focused on facility provision for human excreta. In addition, the uncollected solid waste as well as the disposal of solid waste mainly in dump sites with limited separation and recycling pose environmental and public health problems (Bhatia and Gurnani, 1996; Holm-Nielsen et al., 2009). Regarding grey water management in urban slums, there has been hardly any undertaking. Yet, it is highly polluted with a COD concentration of 2000 mg/L to more than 6000 mg/L and *E. coli* concentrations of 10^5 to 10^8 cfu/100 mL (Carden et al., 2007; Sall and Takahashi, 2006). This PhD study was therefore carried out to contribute to the improvement of sanitation in urban slums with focus on sanitation technologies within the framework of the interdisciplinary research project SCUSA (Sanitation Crisis in Unsewered Slum Areas in African mega-cities). Bwaise III in Kampala (Uganda) was chosen as the study area because it has typical characteristics of urban slums including high population density, poor urban infrastructure and low sanitation service levels (Figure 9.1).

A   B

Figure 9.1:   **A) A view of Bwaise III, the study area and (B) A solid waste dump site next near the poorly constructed housing units in Bwaise III**

Figure 9.2 shows the approach to the study based on an overview of the chapters in this thesis.

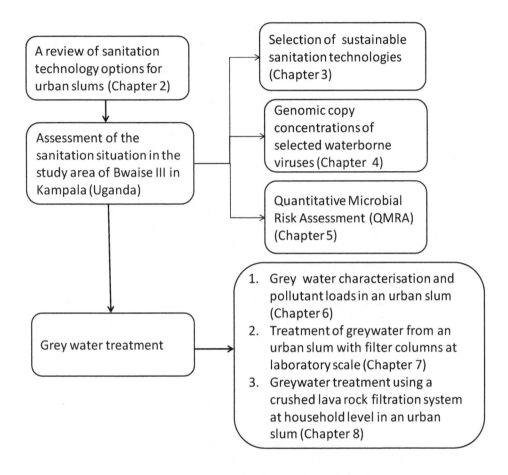

Figure 9.2:   **Overview of the thesis structure**

A baseline study was first carried out in the study area of Bwaise III slum in Kampala (Uganda) to assess the sanitation situation with the aim of selecting suitable sanitation technologies (Chapter 3). The existing sanitation facilities were found unimproved and there was wide spread viral and bacterial contamination in the area (Chapters 4 and 5). In addition, the results from quantitative microbial risk assessment (QMRA) showed that surface water in the open storm water drains, grey water and water storage containers accounted for 39, 24% and 22% respectively, of the disease burden in Bwaise III (Chapter 5). Under non-flooding conditions, the discharge in the open storm water drains originated mainly from grey water. Reducing the grey water pollution load was therefore considered to be essential to maximise the health benefits and also minimise surface water pollution. The study then focused on quantifying the grey water pollution load (Chapter 6) and optimising a media (sand, crushed lava rock) based filtration system at laboratory scale (Chapter 7) followed by implementation of a two-step filtration system at household level in the study area (Chapter 8). In practice, the sustainability and functionality of the crushed lava rock filter to treat grey water in urban slums also depends on the costs, social aspects and the existing institutional arrangements.

9.2   Selection of sustainable sanitation technologies

The high rate of urbanisation in developing countries poses a challenge to urban planning authorities involved in infrastructure development on the one hand (Cairncross, 2006; WHO and UNICEF, 2012) but it is also an opportunity to use new infrastructure investments for resilient and appropriate sanitation systems for entire urban population including the urban poor. Selection of sustainable and technically feasible sanitation technologies is important to achieve the desired public health and environment impacts. Understanding the drivers and barriers of certain technologies requires involvement of the stakeholders that include private practitioners, local authorities, the beneficiary community and policy makers (Mels et al., 2007). The drivers in urban slums include the need to protect the environment and to safeguard public health, resource recovery and existing sanitation policies, legal and regulatory framework, while the barriers include cultural norms, capital and operational costs, resistance to adoption of new technologies and availability of a reliable piped water supply in case of waterborne sanitation systems (Katukiza et al., 2012; Mara et al., 2007; Mels et al., 2007).

Sanitation technologies for urban slums can only be sustainable if they meet the technical criteria for a given environment, comply with environmental requirements and also meet the principles of sustainable sanitation. The principles of sustainable sanitation apply to the choice of technology in general within a sanitation system and also to the specific components of the selected technology. Mara et al. (2007) highlighted the importance of considering all sanitation options at the planning stage in the process of sustainable sanitation planning. In addition, sanitation systems meet the needs of the beneficiaries when they get involved in the planning and design stages (Jenkins and Sugden, 2006). Involvement of different groups of stakeholders also influences the choice of sanitation technologies (Katukiza et al., 2010).

A technology selection method developed in this study (Katukiza et al., 2010) allows making a choice by combining technological and social sustainability aspects during planning for sustainable sanitation in urban slums. The stakeholder perceptions of the existing sanitation technologies were found to influence the outcome of the technology selection process (Figure 3.8, Chapter 3). Providing adequate information about the merits and demerits of all sanitation technology options at the planning stage to stakeholders using Information, Education and Communication (IEC) materials is therefore key in facilitating the participatory approach for sustainable sanitation planning. Analysis of the stakeholders' needs and appreciation of their objectives is essential to identify potential impacts (both positive and negative) and remedial measures. It is also important to accommodate the views of the stake holders in the planning and design stages for the selected technologies. This creates a sense of ownership of the sanitation facilities once they commissioned into operation.

The existing wide range of technologies for managing domestic waste streams (wastewater, excreta and solid waste) are complementary elements within a sanitation system. Therefore, using the technology selection process allows technical solutions to be tailored to the physical context and the demand of the beneficiaries. It may also promote innovations in the sanitation chain aimed at making sanitation provisions in urban slums sustainable. The technology selection method developed under this PhD study is useful for local authorities, practitioners and non-governmental organisations in the urban water and sanitation provision in developing countries. In particular, this technology selection method will be useful for on-going projects that include the economic and demand-led solutions for sustainable sanitation services in poor urban settlements in Kampala, Uganda and the faecal sludge project (The FaME Project) funded by the European Water Initiative ERA-NET (SPLASH) (http://www.splash-era.net/downloads/newsletter_1112.pdf) and a planned study funded by the Austrian development agency to develop a manual for selection of sanitation technologies in small towns in Uganda. It provides guidance in selecting the most sustainable technical solution in an urban slum irrespective of the location. The final decision on implementing one of the most highly ranked options depends on the available funds and the existing policies such as promotion of new innovative technologies.

9.3   Quantification of microbial risks in Bwaise III

A quantitative microbial risk assessment (QMRA) in urban slums is essential to identify and determine the magnitude of a risk to public health from pathogens so that it can be controlled by elimination or reduction to an acceptable level. The microbial risk to human health in urban slums where sanitation infrastructure is very poor is a major public health issue (Figure 9.3). There is very limited information on the magnitude of these risks based on measured concentrations of pathogens especially waterborne viruses through various exposure pathways in urban slums. In this study, the QMRA approach was used to estimate the probable risk of the infection of the exposed population and to quantify the microbial risks from pathogen exposure in the Bwaise III slum in Kampala (Uganda). The reference microorganisms and pathogens were *E.*

coli, Salmonella spp., rotavirus and human adenoviruses F and G. They occur in urban slum environments such as Bwaise III and represent major groups of pathogens (bacteria and viruses) that cause diseases such as diarrhoea and gastroenteritis in areas with inadequate sanitation.

A   B

Figure 9.3:   **A) Grey water from a bathroom stagnant beneath a water meter for a piped water supply house connection in Bwaise III and B) Evidence of exposure to contaminated and unprotected drinking water in Bwaise III**

In this study, the measured concentration of viruses was taken as the actual concentration of pathogens because free nucleic acid (RNA or DNA) is unstable in the environment and infectivity depends on whether the viral capsid is damaged or not for both enveloped and non-enveloped viruses (Templeton et al., 2008; Rodríguez et al., 2009). In addition, the measured concentration of *E. coli* and *Salmonella* spp. were multiplied by a factor of 0.08 to obtain the concentration of the pathogenic strains for use in determining the mean estimated risk of infections (Haas et al., 1999; Machdar et al., 2013). In the absence of data of the actual concentration of pathogens such as rotavirus and hepatitis, the concentration ratio of *E. coli* to pathogen from the literature is used although this may cause uncertainty in the estimated risk of infection.

The sources of contamination used for QMRA in this study included spring water, tap water, surface water, grey water and contaminated soil samples. The actual *E. coli* to pathogen ratio of environmental samples from these sources was determined and compared with values in the literature. The average concentration ratio of *E. coli* to HAdv for both grey water and surface water samples obtained in this study was comparable to the concentration ratio of *E. coli* to viruses ranging from 2.6×10^5:1 to 2.2×10^7:1 obtained in previous studies (Lodder and de Roda Husman, 2005; van Lieverloo et al., 2007) in different environments. The *E. coli* to HAdv ratio of 3.9×10^3:1 obtained in this study was lower than the value of 10^5:1 used by Labite et al. (2010) to estimate the disease burden associated with drinking water treatment and distribution in an informal settlement of Accra (Ghana). The *E. coli* to pathogen ratio varies with time and the level of environmental pollution. The *E. coli* to pathogen ratio from the literature can thus only be used for global estimations of the disease burden, but not for specific urban slums to minimise the level of uncertainty in risk estimation. In addition, the *E. coli* to pathogen ratio may change over time because *E. coli* originates from the entire population every day, whereas the viral pathogens originate from the proportion of the population that is infected at a point in time. In addition, the prevalence of viral infections varies with seasons (Armah et al., 1994). Therefore, there is a need to determine the seasonal variation of pathogen concentrations and loads to increase the reliability of the mean estimated risk of infection.

The total disease burden of 680 disability-adjusted life years (DALYs) per 1000 persons per year was an indication of time lost as a result of the polluted environment in Bwaise III. The highest disease burden contribution in Bwaise III was caused by exposure to surface water open drainage channels (39%) followed by exposure to grey water in tertiary drains (24%) and storage containers (22%). During the dry season, the discharge in the open storm water drains originated mainly from grey water. Grey water was therefore identified as a major source of pollution in the study area. In addition, the results show that the disease burden from each of the exposure routes in the Bwaise III slum except tap water is higher than the WHO tolerable risk of 1×10^{-6} DALYs per person per year. Interventions are therefore needed to improve the health status of the inhabitants and reduce the time lost due to morbidity and mortality. In this study, it was assumed that the microbial risks from dermal contact and inhalation were minor relative to exposure through ingestion. In addition, exposure to contaminated food sources was not considered in our study because the residents obtain food from outside the study area. These assumptions may have caused an underestimation of the disease burden in the study area. There is also a need to obtain epidemiology data to account for uncertainty in the quantified microbial risks in addition to use of Monte Carlo simulation.

Intervention options to reduce the disease burden in Bwaise III were proposed after evaluation of their health effects (section 5.4, Chapter 5). Exposure to contaminated surface water in drains may be reduced by providing a perimeter fence around all drains with provisional gates for operation and maintenance purposes. In addition, storm water drains can be lined and provided with reinforced concrete covers that can only be removed during routine maintenance by personnel with protective clothing. Microbial risks from exposure to grey water may also be controlled by treatment of grey water at household level using media (sand, crushed lava rock, gravel). Grey water treatment at households has an additional advantage of reducing microbial risks from exposure to tertiary drains where untreated grey water is commonly discharged and pollutes surface water. Generally, a combination of the above measures with risk communication through hygiene awareness campaigns is required to achieve the desired health outcomes in urban slums. The final decision for any intervention depends on the financial implications and public health benefits. The QMRA approach therefore, provides support to local authorities in making decisions on the measures to reduce the disease burden to achieve the desired health impacts. In addition, the findings from the quantitative microbial risk also provide guidance and a basis to international organisations such as the World Health Organisation to make strategic investments within the scope of their water, sanitation and hygiene (WASH) activities to prevent disease outbreaks than intervening whenever there is a disease outbreak as has been the case.

9.4   Grey water pollution load Bwaise III

The amount of grey water generated in Bwaise III (the study area) was equivalent to 85±7% of the water consumption and thus accounts for the largest volumetric flux of wastewater in the area. This is typical of grey water production in urban slums (Carden et al., 2007; Morel and Diener, 2006). The quality of grey water from Bwaise III slum was also comparable to that from urban slums in other parts of the world (Carden et al., 2007; Sall and Takahashi, 2006). The BOD_5 to COD ratio of the grey water from Bwaise III ranging from 0.24 to 0.34 was an indication of its low biodegradability. This may be attributed to many reuse cycles, especially for laundry and to a lesser extent household chemicals. The specific grey water COD load ranged from 30 to 75 g.capita^{-1}.d^{-1}, while the COD concentration (> 4000 mgL^{-1}) was higher than the discharge standard of 100 mgL^{-1} set by the National Environmental Management Authority (NEMA) in Uganda. This is likely the cause of the hypoxic conditions in the surface water drainage channels in Bwaise III. In addition, nutrient loads (8.2 kg.day^{-1} of TKN and 1.4 kg.day^{-1} of TP) may cause eutrophication in the receiving water bodies downstream of the slum (Nyenje et al., 2010; Chuai et al., 2013) and ground water contamination through leaching because of the high water table (less than 1 m). Grey water especially from the kitchen had an oil and grease concentration (> 20 mg/L) exceeding the maximum discharge standard for wastewater effluents of 10 mgL^{-1} in Uganda. Therefore, the practice of grey water disposal in the open in urban slums like Bwaise III without treatment may reduce soil aeration and also cause soil infiltration problems in the area. Grey water from Bwaise III had SAR values greater than 6 (Figure 6.3A, Chapter 6), which poses long-term effects of salinity, low water infiltration rates and water clogging in the Bwaise III slum.

Grey water at households and in tertiary drains was polluted with E. coli, Salmonella spp., rotavirus and human adenovirus (Tables 5.3-4, Chapter 5 and Figure 6.6, Chapter 6). The absence of indicator organisms in grey water samples did not reflect the distribution of pathogens in grey water. Waterborne viruses were detected in grey water samples from households and tertiary drains that tested negative for E. coli (Katukiza et al., 2013). There is thus a need for the use of other pathogen detection techniques like quantitative polymerase chain reaction (qPCR) to obtain more reliable data on the pathogen occurrence and concentration. The contribution of grey water to the total disease burden of 24% (which was equivalent to 1.3 x10^{-1} DALYs), shows that grey water poses a risk to public health and contributes to the loss of working days and its management in urban slums needs to be prioritised.

9.5 Decentralised grey water treatment with a low-technology system

Grey water treatment in urban slums is driven by the need to reduce environmental pollution and to minimise risks to human health from the grey water pollution load. A number of low-cost grey water technologies that include horizontal and vertical flow constructed wetlands, grey water tower gardens, and infiltration trenches have been applied in many parts of the world (Li et al., 2009; Masi et al., 2010; Morel and Diener, 2006). Their performance is however, influenced by the grey water quality and quantity. There is thus a need to adapt available technologies to treat high strength of grey water (COD concentration above 3000 mg/L) at pollution sources (households) in urban slums for effective control of the grey water pollution load.

In this study, filter column experiments were carried out (Figure 9.4) to provide a basis for household treatment grey water with a media (sand, crushed lava rock) filter system in an urban slum setting (Chapter 7). The removal efficiencies of COD above 80% were achieved at a hydraulic loading rate (HLR) of 20 cm/day and 40 cm/day while operating a series of two media (crushed lava rock, silica sand and granular activated carbon) filter columns. The removal efficiency of NH_4^+-N, NO_3^--N and TKN at a HLR of 20 cm/day of 79%, 96%, 69% ,respectively, and the log removal of *E. coli*, *Salmonella* ssp. and total coliforms amounting to 3.68, 3.50 and 3.95 at a HLR of 20 cm/day, respectively, showed that media filters have the potential for grey water treatment. It was found that operating the two columns in series at HLR of 20 cm/day resulted in a better effluent quality than at HLR of 40 cm/day. The pollutant removal efficiencies of the filter columns were comparable to values obtained in previous studies (Table 9.1). In practice, grey water pre-treatment followed by a two-step filtration process is therefore required. During filter column experiments, it was observed that filtration through the crushed lava rock resulted in higher pollutant removal efficiency and backwashing was once in two months compared to once in a month for silica sand (Chapter 7). Under field conditions, the clogging effects will be minimised by the wetting and drying cycles as a result of the intermittent flow conditions based on household grey water production.

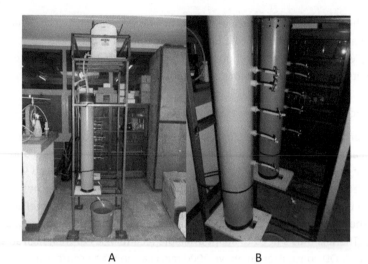

A   B

Figure 9.4:   (A) Filter column set up and (B) location of sampling points on the two
identical filter columns that were operated in parallel and in series

Table 9.1: Comparison of the pollutant removal efficiencies by filter columns with values reported in the literature

Water water type (influent)	Parameter (mg/l)				Reactor					Removal efficiency (%)				
	COD	TKN	TP	TSS	Length (cm)	Inner Diameter (cm)	L:H	Media	HLR (cm/d)	COD	TKN	TP	TSS	
Synthetic grey water	890±130	75±10[a]	4.2±0.2		60	16	3.75	sand	3.2	75±2	5±7	78±7		Sahar et al. (2012)
								bark		74±12	91±8	98±1		
								charcoal		94±2	19±9	92±2		
Synthetic wastewater					30	28.5	1.05	sand and gravel	1.31					Nemade et al. (2009)
Olive mill wastewater	65	2.24	0.42	6.1	100	40	2.5	sand	4	70-90	84-91	nd	75-90	Achak et al. (2009)
un settled sewage	312-410	60-56[a]		274-428[b]	100	7	14.29		2.5					Lens et al. (1994)
								bark	10	72	35		72[b]	
								peat		50	38[a]		91[b]	
Grey water from an urban slum	2861±315	58.5±9.8	2.9±0.5	996±317	60	14.6	4.11	silica sand and crushed lava rock	20	70±4	43±8	52±2	86±2	This study
									40	69±3	39±5	49±3	79±5	
								crushed lava rock	20	70±2	51±2	51±3	85±3	
									40	70±5	42±7	51±2	80±3	

[a] Total Nitrogen
[b] Suspended solids

A prototype crushed lava rock filter based on a two-step filtration process was therefore, designed and implemented at household level in Bwaise III in Kampala (Uganda) to investigate its potential to reduce the grey water pollutant loads (Figure 9.5). The COD and TSS removal efficiency by the crushed lava rock filter unit implemented at household level varied from 75% to 95%. Using an average specific COD load of 55 g COD.capita^{-1}.d^{-1} for Bwaise III, the reduction in the COD load discharged into the environment by an average household in Bwaise III is equivalent to 290 g.d^{-1} based on a 90% grey water collection efficiency. The corresponding reduction in the TKN and TP load discharged into the environment by an average household in Bwaise III is 2.9 g.d^{-1} and 0.4 g.d^{-1} based on removal efficiencies of 69% and 58%, respectively. A log removal of *E. coli, Salmonella* ssp. and total coliforms above 3 was obtained with the crushed lava rock filter implemented at household level. However, the grey water effluent quality did not meet the WHO guidelines for unrestricted reuse. In addition, this study did not cover the removal of viruses during grey water treatment and requires further investigation. Another issue of health concern that requires further investigations is the occurrence of micro-pollutants in grey water. These originate from household chemicals and pharmaceuticals (Hernández-Leal et al., 2011; Larsen et al., 2009.

Figure 9.5:   **A crushed lava rock filter for grey water treatment by a household in Bwaise III**

The filter unit effluent could be further treated in a tertiary step to recover nutrients and remove microorganisms by use low-cost technologies such as the grey water tower garden on which vegetables are grown for consumption (Kulabako et al., 2011; Morel and Diener, 2006). This would increase the removal efficiency of nutrients and microorganisms to produce better quality effluent. The final effluent can be discharged into nearby drains, while the solids after grey water pre-treatment may be dewatered and dried under the sun (feasible in tropics) for pathogen removal prior to use as a soil conditioner. The removal of helminth eggs and other pathogens such as waterborne viruses from the remaining solids by dewatering on a drying bed was not addressed under this study but requires further investigation. The sustainability of such low-technology solutions requires active participation of the household members in order to ensure high grey water collection efficiency and also to maintain the filter performance through routine maintenance. Poorly operated filters are likely to produce low quality effluent and bad odours, which may lead to the households abandoning the use of the filter. The challenge of filter vandalism and theft of valuable filter unit components may be overcome by either securing the filter unit structure, which increases the cost of materials or by using less valuable materials which might shorten the filter's technical life span.

Decentralised treatment of grey water using low-technology solutions that occupy less space (< 1.5 m^2) is a feasible option in an urban slum with limited space and where the ability to pay for services is low. In addition, the negative effects from the grey water pollution load require solutions at the grey water sources (households). Effective management of the grey water stream leads to improved sanitary conditions and thus contributes to an improved health status and quality of life of a slum dweller (Isunju et al., 2011; Okun, 1998). Moreover, grey water was identified as a major contributor to the disease burden in urban slums. This action research demonstrated that low-technology grey water treatment at household level in slums reduces the grey water pollutant loads by 50% to 85%. The wide application of the crushed lava-rock filters at household level in urban slums may provide a better understanding of the impact of onsite grey water treatment on public health and the environment. In addition, the sustainable sanitation systems for management of other waste streams of excreta and solid waste need to be put in place in order to achieve the desired health outcomes in urban slums. This may be achieved through the approach of separating waste streams to facilitate waste treatment and resource recovery. The findings from grey water treatment using low-technology under this PhD study will benefit a new project (with one year duration and funding from UNESCO-IHE Partnership Fund) on assessing the effect of grey water treatment on the sanitary conditions in an urban slum and downstream eutrophication.

9.6   Conclusions

The following are the main conclusions from this research:

- There is no functional sanitation system in Bwaise III. Excreta disposal facilities function as isolated elements and there are no institutional arrangements in place to facilitate collection and treatment of faecal sludge. Grey water is neither collected nor treated. This is typical of the sanitation situation in urban slums in sub-Saharan Africa.

- Grey water generated in Bwaise III may be categorised as high strength wastewater due to its COD concentration above 4000 mg/L. In addition, its bacteriological quality was comparable to that of raw sewage, i.e. the high numbers of *E. coli* and *Salmonella* spp. of up to $1x10^7$ cfu/100 mL and $1x10^5$ cfu/100 mL, respectively.

- The highest contributions to the disease burden in Bwaise III were contaminated surface water, grey water and drinking water storage containers accounting for 39%, 24% and 22% of the total disease burden, respectively. Exposure to grey water in urban slums thus poses a potential risk to human health.

- Grey water pre-treatment followed by a two-step filtration process using crushed lava rock medium reduced the COD and TSS load by over 80% under intermittent flow depending on the household grey water production. A log removal of 3 for *E. coli, Salmonella* spp. and total coliforms was also achieved but the effluent did not meet the WHO requirement for unrestricted reuse such as irrigation. The reduction of the grey water pollution load by the crushed lava rock based filtration unit contributes to public health safety although its sustainability within a sanitation system is affected by institutional and socio-economic factors. The management systems for other waste streams of excreta and solid waste need to be in place to achieve the desired health impacts in urban slums.

9.7   Recommendations for further research

This PhD study has provided an insight of the magnitude microbial risks in slums, and scientific basis for grey water treatment in urban slums followed by application of a low-technology grey water filter unit at household level in an urban slum. The following recommendations for further research are made, based on the findings:

- Integration of quantitative microbial risk assessment (QMRA) in the technology selection for sustainable sanitation options is recommended for future studies aimed at providing a holistic approach for upgrading slum sanitation.

- It is recommended to carry out an intervention and to quantify the health risks before and after implementation and use of the selected technology, especially on the effect of wider application of grey water filter units on the sanitary conditions in an urban slum and downstream eutrophication. This will help to further understand the health impacts and benefits of sanitation solutions, in slums such as the crushed lava rock filter.

- There is a need to investigate the feasibility of adding a tertiary treatment step to the crushed lava rock filtration system investigated in this study to assess advanced pathogen and micro-pollutant removal and reuse of the grey water effluent for productive uses such as irrigation warrants further research.

References

Al-Hamaiedeh, H., M. Bino, M., 2010. Effect of treated grey water reuse in irrigation on soil and plants. Desalination 256, 115–119.

Armah, G.E, Mingle, J.A., Dodoo, A.K., Anyanful, A., Antwi, R., Commey, J., Nkrumah, F.K., 1994. Seasonality of rotavirus infection in Ghana. Annals of Tropical Paediatrics 14(3), 223-9.

Bhatia, M.S., Gurnani, P.T., 1996. Urban waste management privatization. Reaching the unreached: Challenge for 21st century. 22nd WEDC conference, New Delhi, India.

Cairncross S., 2006. Sanitation and water supply: practical lessons from the decade. UNDP – World Bank Water and Sanitation Program, The International Bank for Reconstruction and Development/The World Bank, Washington DC.

Carden, K., Armitage, N., Winter, K., Sichone, O., Rivett, U., Kahonde, J., 2007. The use and disposal of greywater in the non-sewered areas of South Africa: Part 1-Quantifying the greywater generated and assessing its quality. Water SA 33, 425-432.

Chuai, X., Chen, X., Yang, L., Zeng, J., Miao, A., Zhao, H., 2013. Effects of climatic changes and anthropogenic activities on Lake Eutrophication in different ecoregions. International Journal of Environmental Science and Technology 9(3), 503-514.

Christova-Boal, D., Eden, R.E, McFarlane, S., 1996. An investigation into grey water reuse for urban residential properties. In: Desalination 106 (1-3), 391-397.

Dalahmeh, S.S., Pell, M., Vinnerås, B., Hylander, L.D., Öborn, I., Håkan Jönsson, H, 2012. Efficiency of bark, activated charcoal, foam and sand filters in reducing pollutants from greywater. Water Air and Soil Pollution 223(7), 3657–3671.

Elliott, M.A., Stauber, C.E., Koksal, F., DiGiano, F.A., Sobsey, M.D., 2008. Reductions of *E. coli*, echovirus type 12 and bacteriophages in an intermittently operated household-scale slow sand filter. Water research 42(10-11), 2662–2670.

Haas, C.N., Rose, J.B., Gerba, C.P., 1999. Quantitative Microbial Risk Assessment, John Wiley & Sons, New York, USA.

Hernández-Leal L, Temmink H, Zeeman G, Buisman CJN. Removal of micropollutants from aerobically treated grey water via ozone and activated carbon. Water Research 2011; 45(9), 2887-2896.

Holm-Nielsen B, Al Seadi T, Oleskowicz-Popiel P. The future of anaerobic digestion and biogas utilization. Bioresource Technology 100(22), 5478-5484

Isunju, J.B., Etajak, S., Mwalwega, B., Kimwaga, R., Atekyereza, P., Bazeyo, W., John C. Ssempebwa, J.C., 2013. Financing of sanitation services in the slums of Kampala and Dar es Salaam. Health 5(4), 783-791.

Katukiza, A.Y., Temanu H, Chung JW, Foppen JWA, Lens PNL., 2013b. Genomic copy concentrations of selected waterborne viruses in a slum environment in Kampala, Uganda. Journal of Water and Health 11(2), 358-369.

Kulabako, N. R., Ssonko, N.K.M, Kinobe, J., 2011. Greywater Characteristics and Reuse in Tower Gardens in Peri-Urban Areas – Experiences of Kawaala, Kampala, Uganda. The Open Environmental Engineering Journal 4, 147-154.

Larsen TA, Alder AC, Eggen RIL, Maurer M, Lienert J. Source separation: will we see a paradigm shift in wastewater handling? Environmental Science and Technology 2009; 43(16), 6121-6125.

Li, F., Wichmann, K., Otterpohl, R., 2009. Review of technological approaches for grey water treatment and reuses. Science of the Total Environment 407(11), 3439-3449.

Lodder, W.J., de Roda Husman, A.M., 2005. Presence of noroviruses and other enteric viruses in sewage and surface waters in the Netherlands. Applied and Environmental Microbiology 71(3), 1453–1461.

Machdar, E., van der Steen, N.P., Raschid-Sally, L., Lens, P.N.L., 2013. Application of Quantitative Microbial Risk Assessment to analyze the public health risk from poor drinking water quality in a low income area in Accra, Ghana. Science of the Total Environment 449, 134–142.

Mandal, D., Labhasetwar, P., Dhone, S., Dubey, A.S., Shinde, G., Satish Wate, S., 2011. Water conservation due to greywater treatment and reuse in urban setting with specific context to developing countries. Resources, Conservation and Recycling 55, 356-361.

Mara, D., Drangert, J.O., Anh, N.V., Tonderski, A., Gulyas, H., Tonderski, K., 2007. Selection of sustainable sanitation arrangements. Water Policy 9, 305-318

Masi, F., El Hamouri, B., Abdel Shafi, H., Baban, A., Ghrabi, A., Regelsberger, M., 2010. Treatment of segregated black/grey domestic wastewater using constructed wetlands in the Mediterranean basin: the zer0-m experience. Water Science and Technology 61(1), 97-105.

Mels, A., van Betuw, W., Braadbaart, O., 2007. Technology selection and comparative performance of source-separating wastewater management systems in Sweden and The Netherlands. Water Science & Technology 56 (5), 77–85.

Morel, A., Diener, S., 2006. Greywater Management in Low and Middle-Income Countries. Review of different treatment systems for households or neighbourhoods. http://www.eawag.ch/forschung/sandec/publikationen/ewm/dl/GW_manageme nt.pdf [Accessed on 3rd January, 2013].

Nyenje, P.M., Foppen, J.W., Uhlenbrook, S., Kulabako, R., Muwanga, A., 2010. Eutrophication and nutrient release in urban areas of sub-Saharan Africa: a review. Science of the Total Environment 408(3), 447-455.

Okun, D.A., 1998. The value of water supply and sanitation in development: an assessment. American Journal of Public Health 78(11), 1463-1467.

Rodríguez, R.A., Pepper, I.L., Gerba, C.P., 2009. Application of PCR-Based Methods To Assess the Infectivity of Enteric Viruses in Environmental Samples. Applied and Environmental Microbiology 75, 297-307.

Sall, O.,Takahashi, Y., 2006. Physical, chemical and biological characteristics ofstored greywater from unsewered suburban Dakar in Senegal. Urban Water Journal 3(3), 153-164.

Templeton, M.R., Andrews, R.C., 2008. Hofmann R. Particle-Associated Viruses in Water: Impacts on Disinfection Processes, Critical Reviews in Environmental Science and Technology 38, 137-164.

van Lieverloo, J.H.M., Mirjam Blokker, E.J., Medema, G., 2007. Quantitative microbial risk assessment of distributed drinking water using faecal indicator incidence and concentrations. Journal of Water and Health 5 (suppl.1), 131-149.

WHO and UNICEF, 2012. Progress on drinking water and sanitation. Joint Monitoring Program Report (JMP). 1211 Geneva 27, Switzerland.

Summary

This PhD study has shown that to achieve sustainable sanitation in urban slums, technology selection criteria are essential to select appropriate technologies for collection, transport and on-site or off-site treatment of separate waste streams. The results from the selection of sanitation technologies in Bwaise III in Kampala (Uganda) showed that stakeholders' perception of the sanitation technology options determined the ranking of the technologies using the sustainability indicators and the outcome of the technology selection process. Moreover, the participation of stakeholders in the planning, design and implementation of sanitation facilities may affect the functionality and hence the long-term sustainability of sanitation systems.

A baseline study was carried out to determine the occurrence of pathogens and the public health implications in Bwaise III slum as a result of poor sanitation. The results provided new insights on the pathogen concentration, the *Escherichia coli* to pathogen ratio and the magnitude of microbial risks from different exposure pathways in a typical urban slum of Bwaise III. The detection of waterborne viruses in 65% of all surface water and grey water samples was an indication of inadequate human excreta management in the area. The highest disease burden contribution was caused by exposure to surface water open drainage channels (39%) followed by exposure to grey water in tertiary drains (24%) and potable water storage containers (22%). In addition, the disease burden from various exposures was 10^2 to 10^5 times higher than the WHO guideline for the tolerable disease burden for drinking water of 1×10^{-6} DALYs per person per year. The risk of infection and the disease burden in Bwaise III may be reduced by sustainable management of human excreta and grey water, coupled with risk communication during hygiene awareness campaigns at household and community levels.

Grey water was a major source of microbial risks in Bwaise III. In addition, the discharge in the open storm water drains originated mainly from grey water under non-flooding conditions. Onsite grey water treatment was therefore essential to maximise the health benefits and also minimise surface water pollution. Grey water characterisation and pollutant loads' estimation was then carried out followed by filter column experiments to optimise the filtration medium of a grey water treatment system. The grey water return factor in Bwaise III was 85% of the water consumption. The specific COD, TKN and TP loads ranged from 30 to 75 g.capita^{-1}.d^{-1}, 0.1 to 0.5 g.capita^{-1}.d^{-1} and 0.01 to 0.1 g. capita^{-1}.d^{-1}, respectively, depending on the grey water source (laundry, kitchen, and bathroom). The loads of *Escherichia coli* and *Salmonella* spp. in grey water from the laundry and bathroom activities were in the order of 10^6 cfu.capita^{-1}.d^{-1} and 10^8 cfu.capita^{-1}.d^{-1}, respectively.

The pollutant removal efficiencies by the filter columns in parallel mode, using crushed lava rock only and a combination of silica sand and crushed lava rock as the filtration media were comparable. However, the rate of clogging of the sand was two times that of crushed lava rock. Grey water pre-treatment to remove total suspended solids, oil and grease followed by a two-step filtration process was required to achieve COD removal efficiency above 85% using filter columns. The highest log removal of *Escherichia coli*, *Salmonella* spp. and total coliforms amounted to 3.68, 3.50 and 3.95, respectively, at a hydraulic loading rate of 20 cm/day when filter columns were operated in series. In addition, the log removal of *E. coli*, *Salmonella* spp. and total coliforms increased with the infiltration depth during grey water treatment with media filtration under all operating conditions. These results provided a basis for field application of a two-step crushed lava rock filter unit at household level in Bwaise III.

This PhD study also demonstrated that grey water treatment at the generation point (households) to reduce the grey water pollutant loads in urban slums is feasible. The crushed lava rock filter reduced the concentration of COD and TSS in grey water by 85 % to 88% at a varying HLR of 0.5 to 1.1 m.d^{-1} depending on the household grey water production. In addition, a log removal of *Escherichia coli*, *Salmonella* spp. and total coliforms of more than 3 (99.9%) was achieved under household filter usage conditions. Grey water pre-treatment was essential to minimise clogging and sustain the filter performance. A wider application of this technology in urban slums is required to reduce the public health risk and environmental pollution caused by grey water. This may also depend on the availability of funds, existing government policies, social and cultural norms. The management of excreta and solid waste using functional sanitation systems is also required as well in order to reduce environmental pollution and to improve the sanitary conditions and health status of urban slum inhabitants.

Samenvatting

Deze doctoraalstudie heeft uitgewezen dat zorgvuldig gekozen selectiecriteria essentieel zijn om geschikte technologieën te identificeren voor het inzamelen, transporteren en verwerken van gescheiden afvalstromen en daarmee duurzame sanitaire voorzieningen in stedelijke sloppenwijken te bewerkstelligen. In Bwaise III in Kampala (Uganda) werden met behulp van deze selectie criteria en andere duurzaamheidsindicatoren geschikte technologieën geïdentificeerd; de perceptie van de stakeholders was vervolgens bepalend voor de uitkomst van het uiteindelijke technologie selectie proces. De participatie van belanghebbenden in planning, ontwerp en implementatie van sanitaire voorzieningen kunnen mede van invloed zijn op de functionaliteit en daarmee de duurzaamheid van sanitaire systemen op lange termijn.

Een basisstudie werd uitgevoerd in de sloppenwijk Bwaise III om de aanwezigheid van pathogenen en de implicaties voor de volksgezondheid als gevolg van slechte hygiëne vast te stellen. De resultaten leidden tot nieuwe inzichten over de pathogenen concentratie, de verhouding tussen *Escherichia coli* en pathogenen en de omvang van het microbiële risico van verschillende blootstelling routes in een typische sloppenwijk. De detectie van watergerelateerde virussen in 65% van alle oppervlakte- en grijswatermonsters geeft aan dat het management van menselijke uitwerpselen in het gebied ontoereikend is. De hoogste bijdrage aan de ziektelast werd veroorzaakt door blootstelling aan open afvoerkanalen voor oppervlaktewater geopend (39%), gevolgd door blootstelling aan grijswater in greppels (24%) en drinkwateropslagtanks (22%). Daarnaast is de ziektelast van verschillende blootstellingen 10^2-10^5 keer hoger dan de WHO richtlijn voor de aanvaardbare ziektelast voor drinkwater van 1×10^{-6} DALYs per persoon per jaar. Zowel het infectierisico als de ziektelast in Bwaise III kan worden verminderd door duurzaam beheer van menselijke uitwerpselen en grijswater in combinatie met risicocommunicatie tijdens hygiënebewustwordingscampagnes op huishoudelijk en gemeenschapsniveau.

Grijswater is een belangrijke bron van microbiële risico's gebleken in Bwaise III. Daarnaast vormde grijswater de grootste stroom in de open kanalen in perioden zonder regen. Het lokaal behandelen van grijswater is de sleutel tot het minimaliseren van gezondheidsrisico's en oppervlaktewatervervuiling. Grijswaterkarakterisering en een schatting van de vuilvracht werden uitgevoerd, gevolgd door filterkolomexperimenten om het beste filtratiemedium te selecteren voor een grijswater behandelsysteem. In Bwaise III wordt 85% van het totale waterverbruik omgezet in grijswater (return factor van 85%). De specifieke CZV, TKN en TP vracht varieerde van 30 - 75 gCZV.capita^{-1}.d^{-1}, 0.1 tot 0.5 gTKN. capita^{-1}.d^{-1} en 0.01 - 0.1 gTP. capita^{-1}.d^{-1}, afhankelijk van de grijswaterbron (waswater van kleding, keuken en badkamer). De vrachten aan *Escherichia coli* en *Salmonella* spp. in kledingwaswater en badkamergrijswater waren in de orde van 10^6 cfu. capita^{-1}.d^{-1} en 10^8 cfu. capita^{-1}.d^{-1}, respectievelijk.

De verwijderingsefficiënties van parallelle filterkolommen - de een met uitsluitend lavasteentjes, de ander met een combinatie van gewassen silica zand en lavasteentjes als filtratiemedia - waren vergelijkbaar. Echter, de mate van verstopping was bij het systeem met zand twee keer zo hoog als bij het systeem met lavasteentjes. Het voorbehandelen van grijswater om zwevende deeltjes, olie en vet te verwijderen gevolgd door een dubbele filtratiestap was nodig om een CZV verwijderingsefficiëntie boven de 85% te bereiken met behulp van filterkolommen. De hoogste logverwijdering van *Escherichia coli*, *Salmonella* spp. en totale coliformen bedroeg 3.68, 3.50 en 3.95 respectievelijk, bij een hydraulische laadsnelheid van 20 cm per dag en het opstellen van de filterkolommen in serie. *E. coli*, *Salmonella* spp. en totale coliformen werden onder alle omstandigheden beter verwijderd naarmate de infiltratiediepte toenam. Deze resultaten leveren een basis voor de toepassing van een tweetraps filterinstallatie met behulp van lavasteentjes op huishoudniveau in Bwaise III.

Daarnaast toonde deze doctoraalstudie aan dat grijswaterbehandeling op het punt van generatie (huishoudens) om de vuilvracht in stedelijke sloppenwijken te verminderen haalbaar is. Het filter vermindert de concentratie van CZV en TSS in grijswater door 85% tot 88% bij variërende hydraulische belasting van 0.5 - 1.1 m.d^{-1} afhankelijk van de productie van grijswater. Daarnaast kan met een huishoudfiltersysteem een *Escherichia coli*, *Salmonella* spp. en totale coliformenverwijdering van meer dan 3 log (99.9%) behaald worden. Voorbehandeling van het grijswater bleek cruciaal om filterverstopping te minimaliseren en zuiveringsprestaties in stand te houden. Om grijswater-gerelateerde risico's voor de volksgezondheid en milieuvervuiling in stedelijke sloppenwijken te verminderen is een uitgebreide implementatie van deze technologie nodig. Of deze implementatie daadwerkelijk mogelijk is, hangt af van de beschikbaarheid van middelen, bestaand overheidsbeleid en de heersende sociale en culturele normen. Dit neemt niet weg dat het beheer van menselijk en vast afval met behulp van functionele sanitaire voorzieningen alsmede het verminderen van de milieuvervuiling ook vereist is om de hygiënische omstandigheden en de gezondheidstoestand van de inwoners van stedelijke sloppenwijken te verbeteren.

Curriculum Vitae

Alex Y. Katukiza was born on 1st July 1976 in Kabale district, Uganda. He obtained a Bachelor of Science with honours (Upper Second) in Civil Engineering obtained from Makerere University in 2000 after which he started working with a private local consulting firm in Uganda. In 2004, he enrolled for a Master of Science degree course at UNESCO-IHE  funded with scholarship from the Netherlands Fellowship Programme (NFP). He carried out his MSc. research on the effect of wastewater quality and process parameters on removal of organic matter during soil aquifer treatment and obtained a MSc. in Municipal Water and Infrastructure specialising in Sanitary Engineering (Average mark 8.3/10). He then registered with the Engineers registration board in Uganda and started working as a freelance Consultant with local and international companies and has worked on a number of projects funded by the European Union, USAID and World Bank. In January 2009, he obtained funding from the Netherlands Ministry of Development Cooperation (DGIS) through the UNESCO-IHE Partnership Research Fund to pursue his PhD research at UNESCO-IHE Institute for Water Education and Wageningen University.

List of scientific publications

Katukiza, A.Y., Ronteltap, M., Niwagaba, C.B., Foppen, J.W.A., Kansiime, F., Lens, P.N.L., 2012. Sustainable sanitation technology options for urban slums. *Biotechnology Advances 30, 964-978.*

Katukiza, A.Y., Ronteltap, M., Niwagaba, C., Kansiime F, Lens PNL., 2010. Selection of sustainable sanitation technologies for urban slums - A case of Bwaise III in Kampala, Uganda. *Science of the Total Environment 409(1), 52-62.*

Katukiza, A.Y., Temanu, H., Chung, J.W., Foppen, J.W.A., Lens, P.N.L., 2013. Genomic copy concentrations of selected waterborne viruses in a slum environment in Kampala, Uganda. *Journal of Water and Health 11(2), 358-369.*

Katukiza, A.Y., Ronteltap M., van der Steen, J.W., Foppen, J.W.A., Lens, P.N.L., 2013. Quantification of microbial risks to human health caused by waterborne viruses and bacteria in an urban slum. *Journal of Applied Microbiology,* DOI: 10.1111/jam.12368. "In Press".

Katukiza, A.Y., Ronteltap, M., Niwagaba, C., Kansiime, F., Lens, P.N.L., 2013. Grey water characterisation and pollution load in an urban slum. *International Journal of Environmental Science and Technology.* "Accepted".

Katukiza, A.Y., Ronteltap, M., Niwagaba, C., Kansiime, F., Lens, P.N.L., 2013. Grey water treatment in urban slums by a filtration system: optimisation of the filtration medium. *Journal of Environmental Management.* "Under review".

Katukiza, A.Y., Ronteltap, M., Niwagaba, C., Kansiime, F., Lens, P.N.L., 2013. A two-step crushed lava rock filter unit for grey water treatment at household level in an urban slum. *Journal of Environmental Management.* "Accepted".

Netherlands Research School for the
Socio-Economic and Natural Sciences of the Environment

C E R T I F I C A T E

The Netherlands Research School for the
Socio-Economic and Natural Sciences of the Environment
(SENSE), declares that

Alex Yasoni Katukiza

born on 1 July 1976 in Kabale, Uganda

has successfully fulfilled all requirements of the
Educational Programme of SENSE.

Delft, 29 November 2013

the Chairman of the SENSE board

Prof. dr. Rik Leemans

the SENSE Director of Education

Dr. Ad van Dommelen

The SENSE Research School has been accredited by the Royal Netherlands Academy of Arts and Sciences (KNAW)

K O N I N K L I J K E N E D E R L A N D S E
A K A D E M I E V A N W E T E N S C H A P P E N

The SENSE Research School declares that Mr. Alex Yasoni Katukiza has successfully fulfilled all requirements of the Educational PhD Programme of SENSE with a work load of 44 ECTS, including the following activities:

<u>SENSE PhD Courses</u>
o Environmental Research in Context
o Research Context Activity: Dissemination of PhD Research findings on Sanitation in Slums: organizing stakeholder meeting in Kampala, Uganda, publishing newspaper article (Uganda Observer) and producing two instructive movies
o Research Methodology I: from topic to proposal
o Risk Assessment
o SENSE Writing Week

<u>Other PhD and Advanced MSc Courses</u>
o Quantitative Microbial Risk Assessment Summer School Course
o Techniques for Writing and Presenting a Scientific Paper
o Sustainable sanitation

<u>Oral Presentations</u>
o *Management of waste streams in urban slums.* PhD Seminar UNESCO-IHE, 6-9 April 2009, Delft, The Netherlands
o *Grey water characterisation.* PhD Seminar UNESCO-IHE, 20-24 September 2010, Delft, The Netherlands
o *Identification of sustainable technical sanitation options for urban slums.* 10th WaterNet/WARFSA/GWP-SA symposium, 28-30 October 2009, Entebbe, Uganda
o *Evaluation of the potential of a multi-media filter for treatment of greywater generated in an urban slum area using uPVC columns.* IWA World Water Congress and Exhibition, 16-21 September 2012, Busan, South Korea

SENSE Coordinator PhD Education

Drs. Serge Stalpers

T - #0392 - 101024 - C42 - 240/170/15 - PB - 9781138015555 - Gloss Lamination